Africa's Joola Shipwreck

Africa's Joola Shipwreck

Causes and Consequences of a Humanitarian Disaster

Karen Samantha Barton

LEXINGTON BOOKS
Lanham • Boulder • New York • London

Published by Lexington Books
An imprint of The Rowman & Littlefield Publishing Group, Inc.
4501 Forbes Boulevard, Suite 200, Lanham, Maryland 20706
www.rowman.com

6 Tinworth Street, London SE11 5AL, United Kingdom

British Library Cataloguing in Publication Information Available

Library of Congress Control Number: 2020946045
ISBN 978-1-4985-8541-5 (cloth)
ISBN 978-1-4985-8542-2 (electronice)

For my mother, Phyllis Marion Thompson,
a survivor

Contents

Foreword

M. Elie Jean and Bernard Diatta

This introspective research and the fight against forgetfulness that this book proposes is critical to the story of Africa. I wish to tell you about the *Joola* shipwreck and the eighteen years of suffering that we, family members of shipwreck victims, continue to endure.

On Thursday, September 26, 2002, some 1,863 people died aboard the *Joola* ferry, a tragedy which caused enormous sadness in our communities, particularly in the southern region known as Casamance where many of the passengers were born. It is quite difficult and perhaps impossible to cover all the aspects of this shipwreck disaster and the apocalypse that followed. Even as I write this foreword, I feel the visceral pain of a navigation catastrophe that robbed us of our happiness. On that night individuals from twelve different countries—including 450 Senegalese students, twenty-seven members of the football team Mamadou Mareme Diallo—perished unnecessarily, left with the sharks in a cold ocean of staggering consequences.

Of the bodies that were retrieved, many were unidentifiable and buried in mass graves across Senegal and The Gambia. Our local cultural and religious customs do not allow for these informal burials either in common graves or oceanic ones. There are certain protocols involved in paying respect to the dead. So it is shocking that in Senegal, the land of *teranga* and welcoming hospitality, so many have been buried in the deep abyss without proper respect.

Only sixty-four people survived the *Joola* wreck, one of whom was pregnant, all of them left to struggle alone in the aggressive Atlantic waters without assistance from the government. In order to save themselves, many of these survivors swam through tens of kilometers of sea before being picked up by *pirogue* fishing boats. Others hung onto the side of the *Joola*'s hull after it capsized. In its wake the *Joola* disaster would leave behind thousands of orphans and a national sense of trauma for those who lived. As a community there are no words to describe what we feel because we are still trapped in bitterness for what our government failed to do. What makes this disaster so catastrophic is the enduring pain of those who lost family members on the boat, and the lack of recourse due to the government's injustices. Nearly twenty years later, we long for that justice.

Our distinguished pastor Monsignor Jacque once said: "We cannot let the sea take so many people who are dear to us and behave as if nothing happened. We must meditate on the abyssal solitude of those we loved so much, the suffering of their bodies, our community's cultural and religious diversity, while climbing on a new terrain of questioning as to what went wrong."

The *Joola* tragedy represents a flagrant violation of human rights. But we will uncover the mysteries surrounding the disaster and hold those people accountable who tried to bury this injustice so we can finally bury our deceased ones. In light of this, I strongly recommend this book, which illustrates in the smallest detail one of the worst maritime catastrophes in the world, its causes, consequences, and management.

M. Elie Jean Bernard Diatta
Secrétaire des Affaires Juridiques de l'ANFVR-Joola
Ingénieur en Finance/Audit
Consultant et Doctorant à Aix-Marseille Université (France)

Acknowledgments

Emily Roderick took on my fledgling project at Lexington Press (Rowman and Littlefield) and Drs. Judith Lakamper and Trevor Crowell kindly saw the work to its completion. For those who read all or parts of this draft and provided constructive feedback and criticism—particularly Casey Miller, Arun Rajakumar, and Dr. Jerry Mitchell—I appreciate your enduring support. Many scholars and scientists provided inspiration for this work—Haidar El Ali, Pat Wiley, Dr. Charles Becker, Patrice Auvrey, Dr. Fallou Ngom, Katie Jost. My Senegalese and American colleagues were also instrumental in executing fieldwork and arranging field interviews—Dr. Moussa Seydi, Issa Mane, Dr. Elie Diatta, Djibril Sene, Mbaye Ngom, Major Alpha Dia, Ousmane Keita, Dr. Jieun Lee, Hamdine Watt, Kate Mericle, and Michael Chenore.

Professor Elie Bernard Diatta of Casamance advised me that there would be no repercussions for taking on a topic as sensitive as the *Joola* disaster in my capacity as an outsider. He assured me that I was family now, and those words provided great comfort. His support, along with that of my husband and son, played a central role in my finishing the book. I am indebted to each of them.

Finally, while I have gone to great lengths to conduct this fieldwork with support from local partners, any errors in this book are my own. With a few exceptions names have been substituted in order to protect the anonymity of individuals involved in this project. My hope is that young people around the world will grow up knowing the name of the *Joola* in the same way they have come to remember the *Titanic*. May Senegal's resilience in the face of the *Joola* tragedy and their fight for justice be an inspiration to all of us.

Abbreviations and Acronyms

ANFV:	National Association of Families of the Victims of the Joola
AFPRC:	Armed Forces Provisional Ruling Council
ECOWAS:	Economic Community of West African States
EEZ:	Exclusive Economic Zone
FAO:	Food and Agricultural Organization
GIS:	Geographic Information Systems
GGWSSI:	Great Green Wall of the Sahara and Sahel Initiative
IUU:	illegal, unreported, and unregulated fishing
MFDC:	Movement of the Democratic Forces of Casamance
NOAA:	National Oceanic and Atmospheric Administration
RORO:	roll on, roll off ferry
UNDP:	United Nations Development Programme
UNESCO:	United Nations Educational, Scientific and Cultural Organization
WACF:	Women and Children First (law of the sea motto)

Introduction

In 2002, the *Joola*, a government-owned Senegalese ferry, capsized seventeen miles off the coast of The Gambia,[1] killing an estimated 1,863 people and leaving just sixty-four survivors, only one of whom was female, and she was pregnant. The *Joola* is recognized as the second-worst non-wartime maritime disaster in world history, yet few people are aware of this event, nor of the impacts it had on Senegambian society (Barton 2017; *New York Times* 2003; Williams 2012). Despite more casualties than the *Titanic*, to date there has been no academic literature devoted to the *Joola*, nor a comprehensive scholarly treatment of how the agricultural and fishing communities of southern Senegal were disproportionately affected by the event. Among those lost were 450 youth, en route to the capital city Dakar, where many young people were pursuing a postsecondary education not available in their home communities (UNESCO 2009).[2] The lack of attention to the *Joola* tragedy is not surprising, given how little the Western media covers the thousands of events each year that take place across sub-Saharan Africa, both positive and negative (Mericle 2016; Oguh 2015; Osunde 1996). The goal of this book is to increase global awareness about the shipwreck and the communities most affected by the disaster through an in-depth geographical and historical field investigation (Figure 0.1).

Every story has a map, and the tale of the *Joola* shipwreck is no exception. In May 2016, I traveled to Senegal with a team of sixteen professors as part of the U.S. Fulbright program, one of the most widely recognized fellowships for teaching and research in the world. In the wake of World War II Senator J. William Fulbright wanted to create a cultural exchange program with the goal of improving intercultural relations, cultural diplomacy, and intercultural competence between the United States and other countries. As a professor based at a regional comprehensive university where research and travel funds

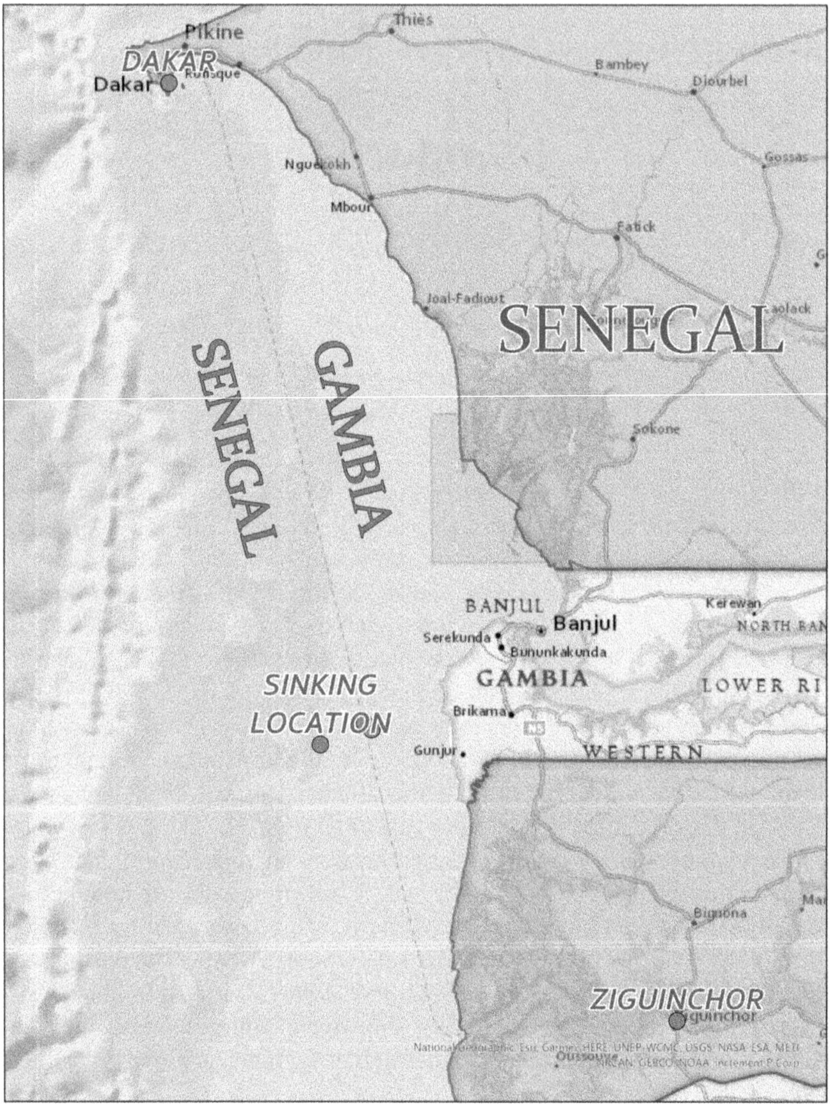

Figure 0.1 Map of Senegal and The Gambia. Cartographer: Dr. Jieun Lee.

are limited, this kind of federal funding provides an enormous opportunity for scholars at our institution.[3] I had worked in other parts of the world courtesy of the Fulbright fellowship—Brazil, UAE, Kuwait, Qatar, Nepal, and soon, Myanmar—but like many participants in this particular program, the grant to Senegal changed the course of my life. I'd never been particularly interested in West Africa, not because it lacked a rich culture but my own biases toward less populated geographic world regions sent me in different trajectories. But as a geographer I was eager to visit Senegal in the heat of the summer African monsoons. I was angling to study the Great Green Wall of the Sahel and Sahara Initiative (GGWSSI), a multinational project to restore acacia-based agricultural systems by planting a wall of trees across the African Sahel.[4] At a time when America was debating the siting of a wall along the U.S.-Mexico border to deter immigrants, the GGWSSI represented the "big idea" that provided a positive ecological spin for combating desertification in one of the areas most impacted by human-induced climate change: Africa.

The goal of the four-week long program was simply to understand religion and diversity in Senegal. To do this, our Boston University hosts—Drs. Fallou Ngom and Tim Longman—had assembled a team of faculty participants from numerous academic fields including history, religion, education, linguistics, art education, with one geographer thrown into the mix. In addition to exploring the success of the GGWSI, I was anxious to understand how a country the size of Colorado—straddling a diverse array of Köppen climate zones from the warm desert, to semi-arid, to tropical savanna—had managed to boast a similar level of cultural diversity despite the imprint of French colonial rule. Here, Wolof, Fula, Seerer, Jola, Mandinka, and Soninke ethic groups live side by side in relative harmony with one another.[5]

These predominant ethnic groups in Senegal share a common cultural background such that apart from their language differences there are few effective barriers between them. On top of this, Senegal is a nation that is 94 percent Muslim, rooted in the Sunni Maliki school of jurisprudence with Sufi influences (Pew Report 2010). Yet despite its Muslim majority status, Senegal's first president was a Christian who served the state from 1960 to 1980. Poet, politician, and cultural theorist, President Léopold Sédar Senghor was among Senegal's minority Christian population of just 4 percent yet he steered the country for twenty years. Today, Senegal remains one of the hallmark examples of peaceful religious coexistence (Pew Report 2010). This fact is particularly remarkable given that Senegal straddles what is often called Africa's "religious fault zone" just south of the Sahara—where Christianity and Muslim majority nations meet—spreading from Mali in the West, to Somalia in the east (Loimeier 2013; Stump 2008). Somehow, the Senegalese—despite the long-standing thorn in their side that is the Casamance conflict[6]—have managed to put religious differences aside to live their lives without faith-based squabbles (Fisher 2012; Pew Report 2010). They coalesce. They converge. Senegalese people make do (Figure 0.2).

I first learned about the *Joola* shipwreck[7] the day before I left Senegal in May 2016. I'd been a geography professor for nearly twenty years, having taught courses on Africa, too—but I'd never heard or read a word about the world's second-worst peacetime maritime disaster. By many it was often

Figure 0.2 Muslims and Christians in Africa. Source: Pew Report 2010.

called "Africa's Titanic," though others would bristle at the depiction. Until that point, I did not know that more than 1,863 people died off the coast of The Gambia on this passenger ferry, more than the legendary *Titanic* at just over 1,500 souls (Biel 2012). Nor did I realize that those onboard the ship were from rural, verdant Casamance in pursuit of better lives in the north, or the ways in which their stories would parallel that of steerage passengers on the *Titanic* who had left Southampton en route to New York in 1906. After all, the *Joola* disaster had happened just one year after America's 9/11, a time when terrorists sent jets soaring into skyscrapers and elsewhere across our nation. I reasoned that perhaps we were distracted by our own history and the *Joola* sank through the media cracks. From Dakar, I emailed friends and family and students back home, and while one colleague said she had recalled reading a news piece about the *Joola* disaster, she was alone in that memory. The reality of September 26, 2002 is that people from fifteen countries, including Gambia, Netherlands, France, Switzerland, Spain, Norway, Cameroon, Ghana, Lebanon, Nigeria, and Guinea, had lost their lives in the wreck, though the overwhelming number of those who perished were from Senegal, most of whom were living in the rural south.[8] In the words of my Senegalese translator, Issa Mane: "Had someone from the USA been on the boat, the *Joola* would've been a Hollywood screenplay by now."[9]

The process of launching the *Joola* project unfolded rather quickly and its creation was a response to requests from Senegalese people who asked me to write the story. One Casamance colleague told me that the Great Green Wall was a wonderful project, but that it was the wrong one, and he and others entreated me to draft the *Joola* book instead. I returned home in June with a new mission in mind—to acquire funding in order to make American audiences aware about the shipwreck tragedy and its human dimensions. In 2017, I requested funds from the National Geographic Society (NGS) so I might return to Senegal for fieldwork, write the manuscript, and produce an interactive GIS (Geographic Information Systems) map that would enliven the story for American audiences, particularly for youth who know very little about the continent and its people (ESRI 2016; Gershmel 2018). To date the NGS has written countless stories about the *Titanic*, rated as the third worst peacetime disaster in terms of lives lost. Likewise, the NGS has produced a documentary film on the *Doña Paz*—the single most severe shipwreck in modern peacetime history, in which the seriously overcrowded ship met its demise in an explosion.[10] By most accounts some 4,386 people perished that day, as the MV *Doña Paz*—a Philippine-registered passenger ferry—sank after colliding with the oil tanker MT *Vector* on December 20, 1987. Yet interestingly the NGS has never covered the African ferry wreck. That the *Joola* story has not been highlighted by the NGS ultimately reinforced the point I had tried to make in regard to the shipwreck: that African stories often go untold, and that Africa in general is undervalued.

In 2017, good news arrived when the Council of American Overseas Research Centers and the National Endowment for the Humanities provided a senior fellowship that would support the book's research and writing. The University of Northern Colorado provided equal support from the Provost Fund, and Dr. Laura Connolly—the Dean of Humanities—afforded teaching release time so that I might conduct four months of fieldwork in country. All told, I would spend three summers in Senegal to explore the geographic roots of the *Joola* disaster. Fieldwork methods involved semi-structured interviews, participant observation, and media and archival research related to the *Joola* shipwreck and Senegal's maritime history.

From the beginning I decided that I did not want to write a disaster narrative. French authors had already created stellar publications that covered the causes of the shipwreck and the demographics that were most affected (Auvray 2012; Becker 2012). Pat Wiley, the only American author to date to cover the topic, had successfully explored in depth the days leading up to the *Joola* tragedy and the event itself through detailed fieldwork and interviews in Senegal (Wiley 2012). I was primarily concerned with relating not just the story of September 26 but unraveling the geographical forces that gave rise to the ferry line's creation in the first place. I wanted to humanize the tragedy and set the story in place—or to underscore the *why* of where. Why did the shipwreck disaster occur in the place that it did, at that particular point in West African history?

Senegal proved to be a fascinating place to work due in part to its extraordinary hospitality known as *teranga*, or welcoming generosity, to strangers. As an environmental geographer I was first drawn to the enormous physical diversity of the country, and the ways in which the semi-arid landscapes of the north were in contrast to the more verdant wetlands of the southern Casamance. Dust storms known as *haboobs* would periodically sweep across the north, including the city of Dakar, signaling a dramatic start to the rainy season. In the southern Casamance, intensive rains would serendipitously strand our team at a compound for the entire afternoon, leading to extended interviews with local families over a plate of *yassa jën* (the national dish of fish and rice). Time spent at mosques, Pentecostal churches and with traditional African healers brought Senegal's history of collaboration and resilience to life. Fishermen and marabouts[11] sat on beaches and explained to me the power of *jinn* and *gris-gris*[12] and the everyday strategies employed to keep their *pirogues* safe while at sea. Later these same fisherfolk invited me and my family to celebrate special events such as the holy Muslim day of Tabaski, where a sheep is slaughtered as part of a feast to commemorate the strength of Ibrahim's faith. We also planted mangrove seeds in coastal wetlands with the former Minister of Ecology Haidar El Ali and his team (Figure 0.3). As a fellow diver and community leader, Haidar would play one of the most prominent and immediate roles in the *Joola* disaster. Whereas the government had failed *the Joola p*assengers, Haidar and local fishermen came to the rescue as first responders.[13] His hands-on efforts in Senegal were not limited to the shipwreck. Haidar has combated coastal erosion and climate change and is a grassroots leader for villages across the country. This is a fairly typical story in Senegal, where Christians and Muslims work side by side to reforest coastal wetlands in the face of global incursions. Theirs is a world made by hand.

Figure 0.3 Citizen Science Team in Casamance. Source: Author.

Everywhere I traveled, locals were ready to discuss the challenges facing their villages and to share these narratives in formal or impromptu settings. Their stories were not just about the *Joola* crisis but extended beyond the tragedy to include discussions about the sea, the divide between the north and south, and the state's failure to incorporate the needs of Casamance communities (Theobold 2015). While Senegalese rural people have enormous grit, they are facing unprecedented changes not just from the state but from larger external forces beyond their control. One example is how artisanal fishermen I interviewed spoke often of the impacts of illegal, unreported, and unregulated (IUU) fishing and the ways in which piracy undermined their food security. Further to the north, the UNESCO world heritage site of Saint Louis market sellers and coastal dwellers mentioned climate change. They talked candidly about how shorelines were eroding, and how their fisheries had been gutted by illegal trawlers. Senegalese made clear that while they wished to document the adverse impacts of commercial fishing and climate change, that my work should shed light on *rafet njort*, a Sufi (the mystic philosophy of Islam) phrase taken to mean great optimism in the face of everyday difficulties (Ngom 2016). It is no understatement to say that Senegalese communities are overwhelmingly resilient in the face of large-scale environmental changes that extend beyond the *Joola* crisis.

Thus the strategy of this book is to explore the tragedy and to make it come alive for the vast majority of those who are unfamiliar with this historic event. But just as important is the need to understand the geographical forces that gave rise to the *Joola's* creation: the south's disenfranchisement from the urban north, the war in Casamance, and the lack of development in rural regions despite their high educational achievements (Fisher 2012). This work also aims to understand Senegal's natural resource base, its complex colonial and political history, dominant views of the sea, and community resilience in the wake of ongoing external threats to Senegalese livelihoods and landscapes. In addition to the events leading up to September 26, 2002, I explore a country and its exceptionalism, so that scholars might better understand the spatial conditions that give rise to, or correlate with, a disaster, while not directly producing them. I situate this humanitarian disaster within the context of larger trends within Africa today (e.g., civil conflict, outmigration, climate change) while concluding with a geography of hope so readers might understand why the *Joola* crisis happened in the first place and where things stand now.

NOTES

1. Gambia is one of only two countries that have "The" attached to its name. Upon its independence in 1965, the nation used the name "The Gambia." This was changed to "Islamic Republic of The Gambia" during Jammeh's reign but on January 29, 2017, the original name was again employed.
2. Approximately 210 of the 450 youth who perished were actually five years of age and under.
3. For more information, see Fulbright Scholar program history. https://www.cies.org/history
4. The Great Green Wall of the Sahara and Sahel Initiative is supported by the African Union, the Food and Agricultural Organization, and thirteen African states.
5. The spelling for each Senegalese ethnic and language group varies depending on the source.
6. The Casamance conflict is an extant dispute between the Senegalese government and the Movement of Democratic Forces of Casamance (MFDC). The Casamance region is ethnically and religiously unique from Senegal, and since 1982 many locals have pushed for independence, arguing that Dakar profits from the south's resource-rich status.
7. This book employs the terms "ferry, boat, and ship" interchangeably in order to align with the multitude of ways in which my interviewees spoke of the disaster. Maritime scholars may disagree with this approach, but it squares with discourse in Casamance.
8. A Senegalese crisis unit reported that foreign passengers included ten French nationals, five Spaniards, two Dutch, two Belgians and two Swiss, along with twenty people from Guinea Bissau. One Frenchman, Patrice Auvray, who resided in Senegal, was among the rescued (*Los Angeles Times* 2002).

9. The author is currently working with colleagues on a film production related to the *Joola* wreck.

With a few minor exceptions real names have been substituted in order to protect the anonymity of those individuals involved in this project. Oral permission for using the entirety of quotes was acquired at the time of the field investigation.

10. *Asia's Titanic, MV Dona Paz* by National Geographic Channel is now available on youtube https://www.youtube.com/watch?v=WAY5XjCLN5w.

11. In Senegal the marabout is often a scholar of the Qur'an or a religious teacher. Others may be wandering holy men who survive on alms, Sufis, or leaders of religious communities.

12. In West Africa *gris-gris* is a term used loosely to refer to a talisman, amulet, charm, or other item or incantation believed to ward off evil and/or bring good luck to an individual or misfortune to another. *Jinn* refers to the spirit world, which is outlined in the "Geographies of the Sea'" chapter.

13. Haidar El Ali can be found in the following E-NCA news video diving the wreck. https://www.youtube.com/watch?v=yuCVB5oNCl8.

Chapter 1

Colonial Cartographies in Senegambia

Whatever is painful does not last, and whoever perseveres smiles.[1]

Wolof proverb

SETTING THE POLITICAL STAGE FOR THE JOOLA DISASTER

Everything is connected to everything else—or in the case of West African geography—everything is not connected which was one source of the problem. The *Joola* ship's very construction was a consequence of Senegambia's fractured geography and the desire to reconnect Senegal as a physically divided nation. While the primary goal of the ferry route was to move people and goods between north and south, the presence of The Gambia as a semi-enclave within Senegal, coupled with the ongoing conflict in Casamance, would ultimately shape the decision to construct a maritime transport line—the *Joola* ferry—linking north and south. This chapter will unravel the political-economic and colonial forces that have structured Senegalese space and transportation over time: the Casamance conflict and its geographic repercussions; and the impact these absurd cartographies have had on maritime and terrestrial patterns. To complete these objectives this chapter and all others employ the use of interview data with local residents in order to best tell the stories and perspectives of those in the region.

Before we understand why Senegalese citizens needed to travel around The Gambia by ferry rather than *through* it, it's important to situate the narrative within the context of the region's specific geographies, and to understand Africa's political history in general. The reality is that following the Berlin

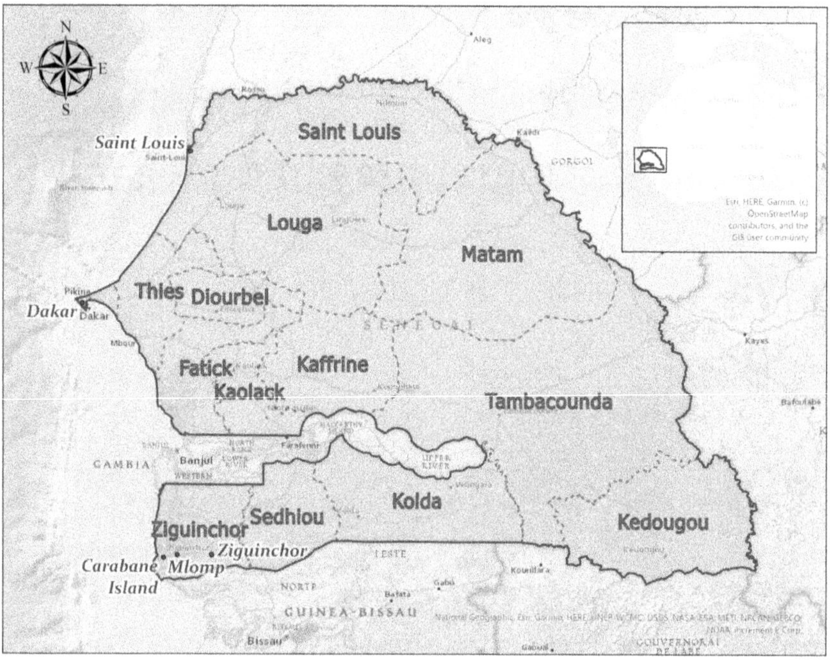

Figure 1.1 Map of Senegal. Cartographer: Dr. Jieun Lee.

Conference of 1854–1855, the West African region known as Senegambia became another casualty of colonial cartography. Sometimes known as the "Congo Conference" or "West Africa Conference," these meetings ushered in a period of heightened colonial activity by European powers, leading to the partitioning of the great African continent exemplified in so many of the political cartoons of the time period (Boas and Dunn 2007). Africa would soon be gobbled up by those most hungry for territory and seized by those least capable of understanding the land and its people. These political geographies would in turn define and delimit transportation routes such as the *Joola* ferry at the close of the twentieth century (Figure 1.1).

THE BERLIN CONFERENCE

The Berlin Conference regulated Europe's colonial endeavors which included trade in Africa, coinciding with German organizer Otto von Bismarck's emergence as an imperialist power in this expansive world realm. While some say the Berlin Conference's role in the partitioning of Africa has been overrated—arguing that numerous bilateral arrangements had been in place prior to this time period—we cannot underestimate the role that the conference played in the "scramble" for Africa. Without a doubt, following the events of 1884–1885, African autonomy and self-governance was eroded, and colonial activity heightened in the wake of the General Act of the Berlin Conference, which permanently reshaped the political geography of the region (Barton 2017). Every major colonial power longed for a piece of Africa, including what is now Senegal, whether they understood its geography or not. As the story of the *Joola* shipwreck will illustrate, little attention was given to cartography's impacts upon oceanic realms.

The Berlin Conference formalized the partition of Africa and made treaties among colonial powers to avoid resource competition. Yet many errors were made in the process due to Europeans' limited knowledge of the continent and the lack of maps in existence. Lord Salisbury, British prime minister in 1906, at the signing of the Anglo-French convention on the border of Nigeria and Niger, famously underscored his own ignorance of geography in stating:

> We (British and French) have been engaged in drawing lines upon maps where no white man's foot ever trod; we have been giving away mountains and rivers and lakes to each other, only hindered by the small impediment that we never knew exactly where the mountains and rivers and lakes were (cited in Anene 1970, 3).

In many cases of partitioning, land was divided without ever having set foot in the Kalahari, the Sahel, the Congo, or what is now the Casamance region of Senegal. There was scant knowledge of the landscape, or corresponding sea territories, let alone the traditional, local communities that founded the great kingdoms ranging from Axum in the East to Timbuktu in the West.

BOUNDARY COMMISSIONS

To be fair, from the 1900s to the 1930s European cartographers conducted surveys of prime African territories through what were called "boundary commissions." The central goal of any boundary commission is simply to "prevent the risk of disputes between national or local populations that arise from overlapping claims to territory or maritime jurisdiction" (Boas and Dunn 2007). In Africa, boundary commissions had the arduous task of applying Europe's political decisions—which defined boundaries—to the local and complex physical and human landscapes of the region. But prior to the twentieth century, these commissions were both bilateral and temporary entities (Gordon and Krech 2012; Klein 1968). In colonial Africa, in other words, there is no historical record of a colonial administration having a permanent domestic or joint commission dedicated to boundary definition and long-term maintenance. It was more common to see joint colonial boundary commissions that were created for a short duration simply to undergo surveys and demarcation work, which would then be disbanded when the colonial mandate was completed. The vast majority of African colonial administrations thus gave little attention to clarifying boundaries or maintaining them on a permanent or regular basis, which would later create issues beyond West Africa alone (Saine 2000).

Many commissions attempted to mark boundaries on the ground, but the way in which they mapped and defined them was inconsistent and not clearly defined. The international land boundaries inherited by African states at the time of independence were the consequence of many different types of inter and intra-colonial boundary commissions that had undertaken survey, mapping, and demarcation work. Yet prior to 1918, most early colonial commissions were not subject to any kind of accurate survey, nor any kind of physical definition or verification on the ground (Tuan 2001). This is what geographers often call "ground truthing." Boundaries between administrative regions of colonial powers were therefore changed or adjusted at will, as was often the case with French administrative "cercles" which affected West and Central Africa. The result was that the definition of international boundaries that were inherited by African states were haphazard, inconsistent, and ambiguous. This critique is especially true for the corresponding sea territories that were the lifeblood of fisherfolk as well as the future "highway" for *Joola* ferry traffic between north and south.

Boundary commissions, by focusing largely on land control rather than people, also ignored the impacts of partitioning on specific and unique ethnic groups. Neither were local *interests c*onsidered. The consequence was what we now see in Senegal and The Gambia but amplified for all of Africa—wherein closely aligned ethnic and language groups were fractured among artificial colonial states. Somalis, for example, were divided into French Somaliland, British Somaliland, Ethiopian Somaliland, as well as the Somali sector of northern Kenya. The Somali people shared a common ethnicity, religion, and culture of everyday life, yet they now live as separate citizens in Ethiopia, Djibouti, and Kenya. This is also true for the Afar people of Ethiopia, who were divided among Ethiopia, Eritrea, and Djibouti, and countless other examples across the continent abound (Dodd 2015).

These artificial borders would create not only cultural barriers but also *spatial* or geographic ones, which in turn hindered peoples' ability to move freely across new and often confusing political barriers. Many Africans such as those in northern Senegal are nomadic pastoralists, and upon demarcating new boundaries, their traditional lifestyles were disrupted. Senegalese or Mauritanian Sahelians who have historically been in need of access to grazing land or scarce water had to navigate not only the challenging physical terrain, but the new political landscape as well. In turn, those in these "borderlands" regions such as the Sahel were relegated to smaller parcels of land, and as such often forced into resource competition with other borderland people who were equally immobilized. It is no exaggeration to say that this inability to move freely caused social problems as well as economic hardships. Structural systems that had been in place prior to colonization were affected and extended families were now separated by a line on someone else's map. There is little evidence of how these new maps affected customary maritime systems in West Africa, and there were few colonial mechanisms in place to police what would ultimately become a twentieth-century *Joola* ferry route.

One of the other consequences in the process of demarcating new boundaries was that Europeans instituted a set of "divide and rule" policies that disrupted the social order and sense of identity for many African communities. Such divide and rule policies set in place new conflicts among local people who at one time may have coalesced. The imperialists instigated conflicts here with the sole purpose of ethnic division, which in turn would serve to strengthen colonial powers (Randall and Coast 2016). In borderlands across Africa today, political elites who affiliate mostly along ethnic lines continue to fuel these age-old but artificial tensions instead of working to alleviate them. The Lou-Nuer of South Sudan and the Jikany-Nuer of Ethiopia are the same ethnic group, both having always resided along the Ethiopia-South Sudan border. However, today they are considered two distinct ethnicities with separate nationalities, and, as such, they compete over limited resources and maintain hostile relationships (Park 2004).

CONTEMPORARY BORDER DISPUTES

As stated, colonial powers set in place these systems but in many ways Africa's political leaders still work to maintain them. Along border areas from Senegal to South Africa, borderlands are now exposed to human security issues that include poverty, limited infrastructure, lower education, and cross-border intergroup conflicts. Today, close to 100 active border disputes exist across the African continent, and with pressure from rising nationalism, population growth, and environmental change, this situation continues to grow. In West Africa alone, modern boundaries and border disputes include (1) land and maritime disputes between Cameroon and Nigeria; (2) territorial disputes on the Island of Mbanie between Gabon and Equatorial Guinea; (3) the frontier dispute between Burkina Faso and Niger; (4) the Benin-Niger frontier dispute; and (5) the Casamance conflict in Senegal (Roche 1985).

In 2007 the African Union stepped in and committed itself to an ambitious border program; it declared an Africa-wide effort to demarcate international land and maritime borders. However, the task itself has been immense, given that the area under consideration amounts to 28,000 miles of international boundaries (Overseas Development Institute 2014). Some efforts—such as the African Union's "Panel of the Wise,"—are capable of addressing these disputes, as are the Council of Elders, which use peace radios[2] and newspapers to address long-standing boundary tensions. These best practices are also evidenced by the work of the Economic Community of West African States (ECOWAS), which has worked to implement ad hoc African arbitral mechanisms and seek the help of neutral groups to address issues.

But while most discussion of African partitioning lays blame on Europe for slicing up the continent, others argue that we must dispel the myth of precolonial geographic harmony (Quinn 1971, 1972). It is clear that Africa was subject to early tensions that resulted in territorial conflicts. In many cases, African cultures relied on the construction of walls and strict boundary markers, which has been evidenced in both oral and written sources. Moreover, recent satellite imagery and archeological research reveals that ancient Africa was reliant upon precise boundary markers that separated states and political groups. For example, there existed approximately 10,000 town walls, 25 percent or more of them on deserted sites, between Lake Chad and the Atlantic Ocean. In addition, evidence of boundaries is found along the 160-kilometer-long Sungbo's Eredo wall, the 45-kilometer-long Orile Owu wall, as well as walls that surrounded full cities including Kwambana, Old Ningi, and Gogoram (Parker and Reid 2013). Still, despite the proof of nationalism and discord between African kingdoms prior to colonization, the key difference is that these walls or borders were African-authored and African-defined. Europeans, who arrived later with cartographers and established boundary commissions, knew far less about the status of borders and the people therein than the kingdoms that once ruled this region. Had these boundaries not been carved in Senegambia, the *Joola* ferry route may have been constructed, but not necessarily required.

THE CASE OF WEST AFRICA AND THE COLONIAL SCRAMBLE

In terms of what is now Senegal, European contact in West Africa and the resulting scramble can be dated as far back as the fifteenth century AD when Portuguese traders made their first links with West African coastal people (Ross 2011). Prior to this time, most Europeans had engaged to a certain extent with West African people by way of the trans-Saharan trade (Ahmida 2009). In the latter part of the fifteenth century the Spanish, Dutch, British, and French all began to establish their presence in the West African region.

Africa, and West Africa in particular, came to represent important possibilities for the expansionist policies of the European powers given its tremendous resource base and ideal geographic position in the eastern Atlantic. The early contacts made by Europeans in the fifteen and early sixteen centuries in the area were mostly focused on trade rather than religious expansion or exploration in general. Here, Europeans traded in slaves, sugar, pepper, ivory, wax, and gold during this period; trade in gold was a central factor in the expansion of European interest in West Africa (though this is not true for Senegal). Gold from West Africa, Ghana in particular, represented 10 percent of the world's reserves in the early part of the sixteenth century and Europe's growing dependence on gold reinforced Europe's links to West Africa (Boahen 1986). Trade is not a static process in West Africa; hundreds of years later, the *Joola* and *Aline Sitoe Diatta* ferries would serve as a regional source of trade, ferrying natural resources from Casamance to Dakar.

The West African coastal peoples encountered by the early European traders represented only a small part of the richness and complexity of the region. Prior to European contact, the region had witnessed the emergence and solidification of a series of ancient African polities, among them, Ghana, Mali, Songhai, and Hausa empires and states. Most notably the Mali Empire (1235–1400)—stretching from what is now Senegal, The Gambia and Mauritania eastward and known for the wealth of its rulers such as Musa Keita—was the largest empire in West Africa (Ngom 2016). It profoundly influenced the culture of the region in terms of its language, laws, and customs. At the empire's height in 1320, Mali covered nearly 1.3 million square kilometers in territory (500,000 square miles), and the kingdom's currency was gold dust (salt, copper, and cowries[3] were also common forms of monetary exchange). The Great Mosque of Djenne, found along the floodplains of the Bani River, represents one of the more iconic images of the Mali Kingdom. Standing at a towering height of 52 feet, the adobe structure highlights Mali's power. It figures into West Africa's precolonial identity and underscores its architectural prowess despite the harsh reality of a desert climate. Together, these precolonial states, Islamic expansion, and a strong precolonial trade network contributed to a diverse social environment into which European traders, explorers, and missionaries entered (Ahmida 2009). Most importantly, it is clear that newcomers underestimated the complexity of the landscape and its people. West Africa—given its resource base and existing social structure—would not be immune from the partitioning but would feature prominently into Europe's legendary scramble for territory.

The contest for what is often called "Senegambia" was fairly straightforward in cartographic terms compared to other parts of the continent. During this time, England declared The Gambia River a British protectorate, and the French, who maintained settlements along the Atlantic coast, were allocated the rest (De Jong and Gasser 2005). As a consequence, Senegal is one of the few modern states to have a political nation partially enclosed within its own territorial borders (Monmonier 2018). Senegal, in many ways, is spatially fractured by the presence of The Gambia. (Though from The Gambia's perspective the optics involve a dominant neighbor who presents its own set of spatial challenges to governance and movement.) Regardless of perspective, few countries around the world must deal with what political geographers call the status of "enclaves," which refers to a territory or part of a territory that is entirely surrounded by that of another state (Blundo 2001). Semi-enclaves—such as in the case of Monaco, Brunei, and The Gambia—can exist as independent states that border only one other state, yet these situations still create intense spatial challenges. These arbitrary colonial divisions in Senegal would serve to split extended families in the region; it would also challenge future transportation networks and contribute to long-term disparities between the north and south of Senegal. There is no doubt that the *Joola* ferry line would be built in response to several of these challenging geographic realities.

CARTOGRAPHY AND ETHNIC IDENTITY IN WEST AFRICA

Cartographies of modern-day West Africa reveal to us the historical connections between ethnic groups that at one time were not split by arbitrary political boundaries. The Mandinka ethnic group is located in both Senegal and The Gambia, an artifact of colonial cartography. Similarly, the Wolof people—a West African ethnic group—are today found in northwestern Senegal, The Gambia, and even southwestern Mauritania, near the coast. In Senegal, the Wolof comprise over 39 percent of the population and are therefore a dominant majority, whereas in The Gambia and Mauritania these ethnicities are minority groups (UNDP 2009). In Senegal, Wolof groups speak the language of Wolof—which is a West Atlantic branch of the Niger-Congo language family. It is possible that Wolof communities in Senegal share more in common culturally with The Gambia or with villages in Mauritania than they do with those in their own country. This problem is not relegated to Senegambia alone. It is common in numerous parts of the world, particularly where culturally similar groups are now divided by arbitrary boundaries, either for political or environmental reasons. Europe's arbitrary postcolonial borders soon left Africans bunched into countries that do not necessarily reflect their heritage, and as we see in the case of Senegal, this contradiction continues to challenge them today. While riding the *Joola's* replacement ferry—the *Aline Sitoe Diatta*—one can hear seven or eight different African languages spoken, evidence of the polyglot population that continues to evolve.

THE GAMBIA AND PRESIDENT JAMMEH

Of all the artificial borders carved up in Africa by colonial powers, one of the more absurd boundaries is the demarcation between Senegal and The Gambia. This particular colonial decision has had long-standing repercussions for the people, politics, and economies of West Africa. In 1964, Africanist Michael Crowder described these impacts in detail, illustrating how the colonial carving created landscapes so diverse despite their common cultural roots. He wrote, in "Colonial Rule in West Africa: Factor for Division or Unity" the following:

> Nowadays it has become almost a platitude to say that the Channel divides West Africa. Nevertheless it is true that when the traveler crosses from Nigeria to Dahomey or from Gambia to Senegal, even at points where the same ethnic group is to be found, he feels that he has crossed those twenty miles of water that separate England and France. It may only be an impression formed from irrelevant details; the long sticks of French bread offered him instead of the oblong blocks of wrapped bread; the change from one side of the road to another, the uniform and equipment of the policeman, particularly the gun he carries in his holster; the cut of the young dandy's clothes; the type of cloth used by women. . . . And the barrier between them is not only of language, but of outlook on basic matters such as law, politics, and culture. The educated Yoruba from Dahomey, in his Parisian cut suit, may have a language and social heritage in common with his cousin from Nigeria, probably wearing agbada; but when it comes to their ideas about democracy and local government, then the real gulf becomes apparent. (Crowder 1968, 44)

As a result of these artificial designations, it is not uncommon to meet Senegalese people who have never traveled through The Gambia (or vice versa) despite its incredibly close proximity to nearly every part of Senegal. It is only a 2.5 hour drive from Ziguinchor to Banjul, The Gambia's headquarters, and just over five hours from Dakar, yet many people I spoke with had never stepped foot next door, despite a strong desire to do so. Over a glass of tea my friend in Ziguinchor shared with me his hesitation to avoid The Gambia, even after the *Joola* wreck:

> One of these days I'll get to The Gambia. But at this point it's too much of a hassle and it's easier to take the boat despite what has happened with the *Joola*.

> The waits are so long, and then sometimes the border just arbitrarily closes. You can sit there for hours in your car having to wait for the next day to cross.[4]

While this problem is challenged by The Gambia's status as a different country, rather just another adjacent county or province, the reality is more complicated than this factor alone. It instead involves economic and political issues that have risen over time (Saine 2000).

For Senegalese, it has simply been difficult to travel through The Gambia, either for reasons related to road delays or the requirements of a political passage into a neighboring state. The West African regional bloc of ECOWAS should ideally allow ease of transportation given its role as a regional economic union in which the "free movement of people and goods" across these states ensues, not unlike the Schengen Accord does in the European Union. But The Gambia presents a far more complicated economic and political situation than this, thus affecting the everyday lives and spatial processes of workers whose livelihood depends upon simply getting key products to market in Dakar or elsewhere. Assuming there are no time delays, it is simply the easiest pathway to travel through The Gambia because it is the shortest route between Casamance and Dakar. However, when the border is closed for a multitude of possible reasons, the 260-mile (420-km) journey *becomes twice as long*. In these instances, travelers or lorries must navigate all the way around The Gambia via Tambacounda or take the boat if the ferry can withstand the weight of their goods. However, in most cases, Senegalese drivers are simply weary of the long road or boat ride when a much simpler, more direct route is available.

Up until quite recently transportation between Senegal and The Gambia was also complicated by the presence of Yayha Abdul-Aziz Jemus Junkung Jammeh, a Jola Muslim from Kanilai who served as the president of The Gambia between 1994–2017 (Saine 2000). It is no exaggeration to say that Jammeh held enormous power in West Africa from his time in the military leading up to his role as president (Dwyer 2017). Jammeh served in the Gambia National Gendarmerie from 1984–1989 until he was commissioned as an officer of the Gambia National Army, where he was able to command the Military Police from 1992-1994. Soon after in 1994, Jammeh spearheaded a bloodless coup d'etat to overthrow The Gambia's government of Dawda Jawara, and in the wake of defeat he installed himself as chairman of the Armed Forces Provisional Ruling Council (1994–1996)—essentially a military junta. He ruled the country by force (Heil 2014).

In his twenty years as president, Jammeh established himself as one of the region's most enduring despots (Obuntan 2015). While The Gambia's economy thrived economically as a result of a burgeoning tourism industry along the coast—not unlike ecotourism in nearby Senegal's Cap Skirring region given its beautiful beaches—the Gambian dictator engaged in wanton disregard for the rule of law and respect for individual liberties. Though a member of the Commonwealth of Nations and thus a signatory to the Harare Convention, Jammeh still ruled The Gambia with an iron fist. He meted out brutal treatment to the Gambia's dissidents and in 2012, ended the state's twenty-seven-year old moratorium on the death penalty, which led to the execution of nine prisoners by firing squad (Abuntan 2015). The Commonwealth government pleaded with Jammeh to show mercy, but he did not waiver; his decision to execute dissidents was especially troubling given that opponents could be jailed for simply questioning Jammeh's promise that he would rule for a "billion" years (Nugent 2007). He also enforced the oppression of anti-government journalists, LGBT communities, and all oppositional parties.

The "Mile 2 prison" in Mandinaring, just outside the capital of Banjul, is a notorious penal institution in The Gambia that is sited on the highway between Serrekenunda and the capital of Banjul. It provides an example of the geographic spaces of oppression instituted during Jammeh's reign. Officially called Central State Prison, it is best known for its grim conditions and the inhumane treatment of and torture of prisoners (Mohamed 2017). Mile 2 is also famous for its treatment of dissidents who did not toe the line of Jammeh's policies, both domestic and foreign. Here, people were imprisoned for opposing the president's registering non-Gambians for voting in the national election. Nokoi Njie, who protested for electoral reform during Jammeh's reign, was imprisoned and tortured yet survived her stay at Mile 2.

> When they put me in a cell, they tied my legs with a wire rope which cut deep into my flesh and I was bleeding. They then beat me with a baton and a rubber whip made from a car tire. They filmed everything they did to me. It was so bad I started saying my final prayers.[5]

Njie was considered lucky, as many political dissidents did not live through this period. Even UN officials, who were permitted inside the country for the first time in 2014, found that "torture is a consistent practice" and that "avoiding arrest is a necessary preoccupation for Gambian citizens" (Archives Nationales du Senegal 2019).

When Jammeh lost the national elections in 2016, yet refused to leave office, he was eventually forced into exile by ECOWAS forces. Adam Barrow was sworn in as the state's new president, and one of his first efforts was to release political prisoners—like Njie—who had been imprisoned without trial. In January 2017, the Senegalese government—in support of new president Adam Barrow—reinforced border checks following Jammeh's hasty departure. These checks around the border led to several arrests, including that of Gambian general Bora Colley, who was caught attempting to enter

Figure 1.2 Gambian President Propaganda. Source: Author.

Guinea-Bissau. General Colley famously served as the commander of the military camp in Jammeh's home village of Kanilai, until he was appointed director of Gambia's prisons (including Two Mile) in 2012. That the dictator had left Gambia was a positive step for the geography of Senegal and The Gambia. However, he had clearly made a mark on transportation routes between these neighboring states. The *Joola* boat had become an established workaround, a clear response to Jammeh's oppressive and long-standing regime in this region.

JAMMEH AND THE MFDC REBEL MOVEMENT

Looking back, the artificial colonial boundaries set between Senegal and The Gambia would serve to divide indigenous communities across the region. But President Jammeh would use this to his advantage in fomenting discord between the two nation-states. Jammeh was both a Muslim and member of the Jola ethnic group of which a small majority can be found in The Gambia (10 percent). The majority of Jola people actually reside in Casamance, Senegal, a region which buts up against the edge of Jammeh's property in Kanilai. Jammeh also worked notoriously to exacerbate tensions between Casamance and Dakar by officially lending support to the adjacent Jola-led Movement for Democratic Forces in Casamance (MFDC). In 2017 Gambia's new government officials reported that Jammeh supporters had been welcoming former rebels into their homes in an attempt to destabilize the Casamance. Residents of Foni, the Gambian region in which the president maintained a vast farm, had hosted MFDC rebels for over thirty years, providing a safe haven in order to upset relations with neighboring Senegal. The new president, Barrow, publicly criticized local leaders for allowing MFDC rebels into their own communities:

> Village and district chiefs are my representatives and are duty-bound to carry out my orders and ensure that there is peace in their places. We are having this meeting because there is a failure in the execution of the task. (Archives Nationales du Senegal, 3)

While the Gambia was seen as an unsafe place for many residents of Casamance, for MFDC rebels, it served as a space for support from the neighboring administration in The Gambia. Yet Jammeh's recognition of the MFDC was not strictly ideological nor related to his ethnic status as a Jola. Some contend that the MDFC movement—which began as fight for independence—increasingly evolved into a racket smuggling timber, drugs, and weapons from which President Jammeh is believed to have personally benefited (Boas and Dunn 2007).

Research shows that the presence of one of the world's great dictators certainly played a key role in the *spatial* relationships between Senegal and The Gambia. At the national level, Senegal-Gambia relationships were often strained, with Senegal silent in terms of Jammeh's transgressions, not willing to add fuel to the fire in an already contested world region. In sharp contrast to The Gambia, Senegal had long been seen on the international level as a relatively peaceful state. Most recently, Senegal has been characterized by the Pew Research Center as a site of peaceful coexistence despite what is a "religious fault" zone of the Sahel belt that spans the length of the continent (Pew Report 2010). Juxtaposed against Jammeh were Senegalese leaders such as Leopold Senghor, the PSS (Socialist Party of Senegal) party leader who served twenty years as president. His faith-based traditions as a Christian, and despite his presence in a Muslim nation, were in sharp contrast to Jammeh's isolationist policies and close-minded ideologies. Abdou Diour (PSS), who served Senegal for nineteen years—followed by Abdoulaye Wade's twelve-year leadership—while fraught with their own set of political problems—were still no match for the perils of Jammeh's reign. It was filled with delusions of grandeur often parlayed into nationwide propaganda (Figure 1.2). Young Gambians risked their lives trying to flee his reign, en route to Europe often by way of *pirogues.* And in the meanwhile, Jammeh not only went after his opponents sending them to Two Mile, but in his fervor claimed to have found an herbal cure for AIDs (Bosha et al. 2007). It would make logical sense that nearby Senegal would want to distance itself politically from his regime despite its proximity and their shared cultural history. It would also make

Figure 1.3 Gambian Border with Senegal. Source: Author

sense why Senegalese wished to avoid even traveling through The Gambia, preferring to hop on the overnight *Joola* instead (Figure 1.3).

THE BRIDGE THROUGH THE GAMBIA

The issue of transportation and politics—whether by land or sea—was consistently mentioned during my *Joola* interviews in Senegal between 2016-2018. It was not part of my research to meet with Gambians, but I suspect that they have similar transport issues complicated by both colonial history and the challenges of two rulemaking states. Interviewees from Casamance frequently mentioned their frustrations with the delays to cross The Gambia. The *Joola* ferry was a logical answer for Senegalese travelers and merchants seeking an easy route to other parts of the country, though this was less true for lorry drivers whose trucks could not be accommodated by the mid-sized ferry. In 2019, some seventeen years after the *Joola* wreck, and nearly twenty years after Jammeh took power in The Gambia, a land-based transport solution finally took form. On January 21, the Senegambia bridge (also known as the Trans-Gambia Bridge) was completed, a 1.9-kilometer (1.2-mile) linkage that forms part of the Trans-Gambia Highway and the Trans-West African Coastal Highway—once and for all linking Senegal and The Gambia—two states that were not altogether different prior to the arrival of the French in Africa.

Funded with a $65 million loan from the African Development Bank and funds from the Gambian government, construction actually began during Jammeh's reign (Sene 2019). The bridge was built by a Senegalese company (Arezki Group) and the Spanish group Isolux Corsan. Made of reinforced concrete, it includes a 100-meter wide central span so that boats may continue to utilize the region for transport under the bridge (Ministry of Equipment and Transport 2002). As mentioned, the ferry crossing would often cause delays, and due to its unreliability, it was not uncommon to wait several days to cross the Gambia River. One Senegalese truck driver, Mawdo Saine (Reuters) said that "A few months ago, I spent ten days at the ferry terminal before it was my time to cross." A colleague of mine in Ziguinchor told a similar story of having to sleep overnight in his car at The Gambian border. Likewise, these issues of long wait times were often confounded by border disputes between Senegal-Gambia. In April 2016, for instance, the border frontier shut down because of a two-month blockade by Senegalese drivers. This happened because the levy they had to pay to cross The Gambia was increased by ten times—from 4,000 CFA per lorry to 40,000 CFA ($700). Time was also an issue in this process. As one lorry driver stated, "There's a lot of difficulty for drivers because what you should do in a week, you must end up doing in one month." While the Senegalese government had not been vocal about these issues for fear of antagonizing its neighbor, the issues hit lorry drivers hard, especially for those who often had to wait months during such crises. The alternative—to take a 250-mile detour around the river—was simply impractical for many, owing to the extra fuel costs. From the Gambian perspective, there are few sources of revenue besides groundnuts and tourism, so travel taxes are a necessity (Gordon and Krech 2012).

The Senegambia bridge is now one of the largest bridges in Africa but it reflects more than superlatives alone; the bridge is politics made manifest in material form. The Gambia's president Adam Barrow and Senegal's president Macky Sall inaugurated the bridge and were the first to cross it in tandem in January 2019. The mayor of Ziguinchor (Abdoulaye Balde) and Senegalese national football team coach (Aliou Cissee) were also present at the event, and thus it was politicized for its attempt to "bridge" these regions as much as it was praised for being an architectural marvel. At the ceremony near the foot of the new structure, Barrow himself stated that "Given the geography of The Gambia and Senegal, with a river dividing each country into two parts, one wonders why it took so long to get this bridge." Indeed, the Senegambia bridge will work to promote trade in West Africa as well as increased spatial interaction. That the bridge was completed after decades of delay symbolizes the renewed understanding between the two countries.

At the time of this writing, the Senegambia bridge has not yet been open to lorry drivers, and while this will take some time, it will eventually have positive impacts for unification in Senegambia, not to mention it will provide economic assistance for those in the marginalized south. Clearly it will be much faster and easier to get products to market in Dakar, and the bridge itself is a symbolic show of unity in a region that has long been at odds. Having said that, any transport decision has spatial consequences for society and space, and there has been some discussion about those river businesses that had become dependent upon financial interactions with lorry drivers who crossed via ferry. Indeed, small vendors would sell food and drink to those crossing the river, and now these groups are grappling with new ways to make a living in the wake of the bridge's construction, which has in some ways displaced their livelihoods.

This discussion reveals how today's transportation patterns in Senegambia—despite the region's small size—are surprisingly complex due to forces which were put in place at the time of colonial partitioning. Colonial cartography laid the foundation for the 2002 *Joola* disaster. Imagine for a moment that the state of Colorado—which is roughly the size of The Gambia and Senegal—had a tiny enclave within it called Denver that operated as a different political entity. Those in larger Colorado spoke English, and those within Denver did not. Denver was overseen by a dictator, and Colorado at large functioned as a democracy. Yet in spite of all this, both Colorado and Denver shared a similar history, laden with indigenous traditions dating back thousands of years. Then imagine you are a truck driver and your livelihood is dependent upon ferrying products from Pueblo, in the south, to Fort Collins in the North. While this route, unobstructed, would usually only take 3.5 hours, at arbitrary times the border with Denver would close, rendering you a refugee for up to weeks at a time. As a geographer friend explained when I shared my metaphor:

> Now what you say is true. But you've left out something important. Imagine you are stuck at a Colorado-Denver border checkpoint, but sitting in a 30-year-old truck, hungry, in 100 degree summer heat.[6]

When seen from this perspective, the challenges involved in traversing Senegambia seem almost absurd, yet Senegambians have had to contend with this issue for generations, made particularly challenging during Jammeh's brutal rule.

While this problem will inevitably change with the arrival of the new Senegambia bridge, it is important to understand the geographic forces that were at work at the time of the *Joola* crisis. Though a seemingly easy area to traverse from the sky, the on-the-ground transport mechanisms for Senegalese, especially those from Casamance, were both limited and complicated, and only served to fuel the *Joola's* eventual creation. This colonialism and the way in which space was structured certainly set in motion the existing networks that would lead to the *Joola* line's establishment, forever altering the lives not only of Senegalese but also of those passengers who chose to board the ship on September 26, 2002.

POLITICAL GEOGRAPHY OF THE MFDC SEPARATIST MOVEMENT IN CASAMANCE

This chapter has outlined ways that colonial patterns created transport and economic barriers for rural people in Casamance that would later disenfranchise and isolate them. Yet it is equally important to discuss how the south's geographic separation from the north (Dakar) at large led to a rebel-based movement in Casamance that would ironically *further* isolate it from the north. I now turn to the role of the Movement of Democratic Forces of Casamance (MFDC)—the main separatist movement in Casamance—in helping to understand the geography of Senegal immediately prior to the *Joola* shipwreck. Founded in 1982 and supported by Guinea-Bissau president Joao Bernardo Vieira until he was overthrown in 1999, the MFDC movement itself was led by Father Augustin Diamacoune Senghor who signed a peace agreement with President Wade in 2004, which remained in place until his death in 2007 (Foucher 2011; Leary 1970, 1971). However, several factions of the MFDC continued to fight despite this civil agreement. The resulting internal division served to deeply divide the Casamance independent movement and affected peoples' ability to move safely within and across the region of the south. The MFDC was a long-drawn out guerrilla war waged for over thirty years until a ceasefire in 2014. Explaining the core reason for the conflict, one of the MFDC's leaders, Youssouf Coly, simply stated that "this conflict is cultural" (Mahling Clark 2011; Marut 2002).

Who are the human actors involved in this separatist movement and what are their grievances? It is well known that the MFDC is a predominantly Jola-based movement. More specifically, the Jola (or Diola, in French transliteration) are an ethnic group found throughout Senegal, Gambia and Guinea-Bissau (Reboussin 1995). In Senegal they predominantly inhabit the region of Casamance, where Jola are believed to have predated Mande and Fula peoples beginning in the riverine coast of Senegambia, migrating into Casamance even prior to the thirteenth century (De Jong 2005). The Jola, for which the *Joola* ferry was named, are actually the main ethnic group in the Casamance, but comprise only a small percentage (4 percent) of the country's total population. As noted earlier, the Wolof are the dominant ethnic group, comprising 43 percent of Senegal (Mahling Clark 2011). Colonial attempts to subjugate the Jola met with little success, since they were known as fierce resisters, and over time, one of the few groups to have kept their culture intact. The Jola were also the last community to accept Islam or Christianity, and today they still maintain animistic traditions that center around a supreme being called Emit, or Ata Emit, the one who resides in the sky (Babou 2007). Senegalese oral history shows us that like many indigenous religions in Africa, nature and the role of sacred forests and wetlands of the Casamance have great importance in their cosmology (Klein 1968).

Many have argued that the south's—and by virtue, the Jola's—economic and geographic marginalization within the larger state apparatus contributed to the founding of the MFDC in 1982 (Deets 2016). Although it was incorporated into French West Africa, many of Casamance's woes have emerged *since* Senegalese independence in the 1960s, where locals felt exploited by Wolof-dominated governments in Dakar. Simmering resentment toward Dakar's influence boiled into an armed conflict in 1982, with separatists seeking independence for Casamance. Since that time, the separatist movement has been fighting for an independent Casamance given its unique needs, resources, and marginalization. The movement began first through popular protest, and then beginning in the 1990s, this insurgency became characterized by sporadic violence. This new development, that of guerrilla warfare, in turn exacerbated existing transportation challenges for those seeking to circumvent the conflict and continue their livelihoods (Deets 2016; Dejong 2005). The war also directly affected subsistence economies, since conflicts with rebels altered the region's summer harvesting activities. For example, the presence of landmines meant that farmers could not work in the fields in northern Casamance safely nor easily move their produce to markets (Barton 2017).

In the same way it is difficult to generalize about the politics of Senegal, it is problematic to look at Casamance as a homogenous political space. Certainly, the people of Casamance are challenged by their isolation in the south, but not all of them have joined the MFDC movement. For example, in Ziguinchor proper (the capital of the south) there is little show of popular support for the rebel movement. According to Nouah Cissee, MFDC historian, "The MFDC only recruits in *villages*" of Casamance, which is in part a function of the dominance of the Jola in the movement (De Jong and Gasser 2005). Yet despite this notion, all Jola people are far from being separatist, but most separatists are in fact Jolas. Residents within Ziguinchor perceive the MDFC separatists as idealists with a shared vision, but at the same time, are troubled by the ways in which their efforts have ironically hindered development. As Amadou Sylla, head of SOS-Casamance stated: "We need to bring the combatants out of the bush and integrate them socially. Some only know about fighting" (Evans 2009). Moreover, the MFDC movement has changed since its origins with the charismatic leader Augustin Diamacoune Senghor. Since his death in Paris in 2017, the MFDC has split and now boasts three armed groups and two secretary generals. Mediators like Mario Giro of Sant'Egidio Catholic community have attempted to intervene with peace talks but they argue that such division within the ranks has challenged things. Senegal has deployed several thousand troops since 2009 to root out the guerilla units, but the MDFC are operating on home turf. Conversely, the MFDC lacks the firepower, and as noted, the support of all Casamance people (Evans 2009; Mahling Clark 2011).

In speaking with many of my colleagues in Senegal, particularly those in Casamance, it was clear that they were conflicted about the MFDC movement. Discussing the MFDC was not recommended, and I was cautioned to set this topic aside. I only discussed the rebel movement when the issue was raised by locals, or from a distance—either outside of Casamance in Dakar or from my home base in Colorado over Skype. One of my colleagues, a professor in the United States, described how he took a more Pan-African view toward politics and space in Casamance that was in contrast to the MFDC approach. He mentioned that globally the trend has been toward regional unity rather than what he called "tiny kingdoms," referring to the goal of an independent Casamance lorded over by the MFDC. The African Union, for example, recognizes eight regional economic communities (UMA, COMESA, CEN-SAD, EAC, ECCAS, ECOWAS, IGAD and SADC). Each of these is closely integrated with the African Union's work and they serve as building blocks for continental policies. And, despite existing challenges with these entities, ECOWAS—which affects Senegal and The Gambia—is among the best-performing economic communities on the continent (*New York Times* 2019).[7] Over *attaya,* a local in Ziguinchor quietly pointed out to me that the world is moving toward such regional cooperation, and the rebel movement in fact subverts these efforts, despite its good intentions. Many in Ziguinchor believed it was possible to promote regional cooperation without sacrificing Casamance's regional uniqueness. Nor did they lay all blame at the feet of the rebel movement, but the Senegalese military, which made its strength visible in public spaces, including its oversight and occupation of the *Joola.*

For the purposes of this book the reality of Africa's "oldest war" is that the everyday movement of people is hindered, which in turn affects the unity of the Casamance region as well as the nation. Examples of the lack of safe travel were evident in interviews during my time in Senegal as well as through my own personal experiences. I was told that I'd "probably be fine" traveling north from Ziguinchor to Bignona, but that "you should be careful in making your way up toward The Gambia border." My friend Thierno would scold me for spending so much time in the field, reminding me that "the roads are dangerous, more dangerous that you think," citing example upon example of the ways in which harm might come to me if I ventured beyond Ziguinchor.

Just one year after my visit to Senegal—where I rented a *pirogue* to view up-close the mangrove forests of Casamance—a bridge had been blown up by rebel forces, an innocent fishermen killed in the process and the nearby local restaurant destroyed (Figure 1.4). In September 2018, just a few months after I'd been in the seaside town of Cap Skirring, I was warned to steer clear of this tourism community since the road to Cap had been frequented by MFDC members. The impact that the MFDC has had—both then and now—on the movement of people across space is critical and is a key to understanding why this region has been so disenfranchised. The MFDC activity reflects the region's dissatisfaction with the north, and its marginalization both prior to and after the *Joola* crisis, but ironically the actions of the MFDC have in some ways hindered growth for the region.

Many tourists are told to avoid Casamance altogether given the some of the dangers that prevail in this tropical paradise. This means that few Americans may ever visit Casamance and experience the marvels of its wetlands and baobab forests. The U.S. State Department in fact warns travelers to be cautious in the Casamance region "due to crime and landmines," while other nations offer more detailed and onerous advice (U.S. State Department 2020). The New Zealand government, for example, regards the security region in the Casamance region between the border area of The Gambia and Guinea-Bissau as "highly unstable" (New Zealand Government 2020). They advise travelers that armed bandits have been known to ambush cars which may

Figure 1.4 Casamance Wetlands. Source: Author.

result in injury and death. The extent of spatial movement is also restricted in that all road travel should be avoided, with the exception of the road from Cap Skirring to Ziguinchor and that of Sao Domingos to Ziguinchor.

> The area to the West of Kolda is particularly dangerous. Any travel that is undertaken should be on well-used roads and only during daylight hours as the threat of banditry increases at night. Landmines are an additional concern in the region, particularly on the Guinea-Bissau border and in Basse-Casamance. As demining operations are ongoing, New Zealanders should seek local advice and stick to paved roads.[8]

Such restrictions clearly impact tourism in Casamance, in some ways adding to its further marginalization. This is highly unfortunate, given the splendor of this region's natural resource base and unique cultural history of cohesion. Cap Skirring was once a popular seaside resort for Europeans, hosting a golf course and an airport built in 1960 by the Aero Club de Ziguinchor. The French, in particular, had recognized it as a balneal zone—a spa town for physical therapy —but such levels of ecotourism have since waned in response to the incursions (Gashaw 2017). While many French tourists would travel on the *Joola* ferry to reach Casamance, the south would receive far less visitors from France in the wake of the shipwreck and the MFDC conflict.

Bayotte Forest Bludgeoning of 2018

Despite a few peaceful years of cease fire, the MFDC conflict reached a climax when on January 7, 2018, in the commune of Boutoupa-Camarcounda—just north of Sedhiou—gunmen attacked and executed thirteen people in the Bourofaye Forest (Bayotte Forest). The gunmen ordered a group of men, including many youth out in search of wood, to lie on the ground of the forest before opening fire upon them. In response President Macky Sall declared the act an "armed attack of rare barbarity" and summoned the Security Council and a ministerial delegation to the scene (Agence France-Presse 2018). During the Bayotte incident thirteen wood harvesters were shot dead, two victims were stabbed to death, and one was burned, with at least half a dozen more wounded, the latter of which were sent to Dakar for treatment. Those who survived described being hit by "two bullets in the back" and fired on by men in "military uniforms, wearing ranger type boots, and speaking the local language" (Agence France-Presse 2018). In response the Senegalese army deployed 150 troops to help evacuate victims and flush out the attackers, with residents in Bourafaye Baniuk reporting that shots were heard as the army pursued a deep search. I had left Casamance in November of 2017, just two months before the incident occurred and was not present to interview my contacts after the massacre. However, a team of Senegalese and North American linguists and anthropologists who were working in the region on an endangered manuscript project—reported hearing explosions near the border with Guinea-Bissau. It was a terrifying time for those connected to Casamance, but much of the backstory that gave rise to this event remained untold in the mainstream press.

While news articles mentioned the forest of Bourofaye as the site of the massacre, the killings actually took place in the commune of Nyassia, which is only twenty kilometers from Ziguinchor proper. There are also several villages that make their home in the Forest of Bayotte, including Boffa Bayotte, Bouhouyoum, Toubacouta, Katoure, Badem, and Baguame. In these villages, locals have for some time witnessed the disappearance of old growth forests—particularly *teakwood*—which has been illegally harvested and shipped to other nations. Teak is a precious wood that is used for boat-making and garden furniture, and villagers in Casamance have become increasingly aware of widespread deforestation in support of global trade. Using community-based tools, these locals have organized politically, creating vigilance committees to prevent woodcutters from Ziguinchor, other parts of Senegal, or even neighboring countries from stealing their wood. Some of this timber has been illegally loaded onto the ferry boat, while the remainder is transported through a network of land-based transport routes both within and beyond Senegal.

Despite these community-based protection efforts in the Bayotte Forest, there have been numerous altercations with illegal woodcutters, some of which have been handled poorly by the state. As recently as October of 2017, for example, members of the inter-villages monitoring committee found a recidivist woodcutter in Bayotte, which the locals presented to authorities (Agence France-Press 2018). The state did not arrest the woodcutter for his theft, but rather the woodcutter lodged a complaint against the villagers, who were soon sentenced by the court and imprisoned for more than one month. These innocent locals were finally released in December 2017, just a month before the Bayotte massacre, which explains in part the roots of this "resource" conflict. Tensions between villagers and woodcutters remain strong, and they are fueled by the power of local authorities such as the governor and the Water and Forest Service. These poor state level decisions have translated into an appropriate local response of vigilantism and regionalism, one that has its roots in Casamance's general discontent with the state.

Many of the news media articles about the Bayotte massacre also described young people simply "out looking for wood" (*News 24* 2018). The image of woodcutters out searching for wood paints a portrait of local people collecting timber for their everyday use without any mention of ownership patterns. While the massacre was a horrific response, its portrayal disconnects the narrative from the larger political-economic structures of extraction taking place within Senegambia. In truth, the history of resource extraction in the Bayotte Forest and other Casamance forest communities is complicated, made more interesting by the fact that these are "protected territories" on the same status with U.S. National Parks. Environmentalist Haidar el Ali has brought attention to this extraction issue as well and highlighted its roots in terms of global resource demand of endangered Casamance timber (Cochrane 2016). The reality is that logging of rare and precious teakwood is forbidden in forests such as Bayotte, yet locals still "look for wood" and then ship it to The Gambia. Thus while Senegalese authorities argue that the massacre was carried out by villagers with rebel support, the MFDC in turn have pointed the finger at an illegal logging trade run by corrupt government officials. While many of the details are nuanced, what is clear is the way in which Casamance space is structured—and the contestations over scarce resources in this marginalized region, have together served to exacerbate long-standing problems.

In the end, many argued that the MFDC rebels were to blame for the massacre. They have implicated the head of the MFDC's southern front citing his frustration with peace talks between MFDC and Senegalese authorities. But the truth is more complex than politics alone. The traffic of teak timber from Bayotte Forest supplies both Ziguinchor timber mills and international trade in The Gambia. Thus the separatists argue that authorities at the state level should focus on the "clandestine" networks of teakwood which enjoy the complicity of the police, local, and regional authorities (Environmental Justice Atlas 2018). The MFDC sees the massacre as a result of rival gangs of woodcutters that are linked to sawmills. Despite these views, the Senegalese army used the opportunity to reinforce its presence in the Bayotte Forest (and Casamance, more generally) by claiming that it planned to neutralize the MFDC rebels and fight the wood traffickers.

The Bayotte Forest narrative has also been portrayed by outsiders as a religious fault zone in which Muslims and Christians are constantly at war with one another. Global news media accounts repeatedly focused not just on the horrific acts of the crime, but proceeded to describe how Casamance is a predominantly Muslim region. While the vast majority of Senegalese are indeed adherents to Islam in various forms, religion has no connection to the Bayotte Forest massacre (Levtzion et al. 2000). In fact, as noted in other parts of this book, the Casamance region is perhaps most known for its peaceful coexistence when it comes to different faith-based traditions (Azuma 2001). Muslims, Christians, and animists may be found sitting side by side on plastic chairs in the neighborhoods of Ziguinchor, drinking *attaya*, attending weddings and participating in various religious festivals such as Tabaski whether they are adherents or not. That the media chose to focus on the religious aspects of the Bayotte Forest massacre gives more credence to the hypothesis that Africans are underestimated, and often subject to gross generalizations about their culture and everyday life. It is akin to a U.S. news report mentioning that an indicted assailant is black or Muslim, which is simply not central to the story itself but only reinforces dominant discourses about cultures and identities.

OUT OF SIGHT, OUT OF MIND

It is important to point out that the MFDC conflict, despite being one of Africa's oldest civil wars, is less well known than other civil conflicts in West Africa such as Liberia and Sierra Leone. The lack of understanding about the MFDC in some ways parallels our ignorance about the *Joola* shipwreck. Part of this is due to the fact that the MFDC crisis is a relatively small scale phenomena (Evans 2002). In Sierra Leone, the number of combatants, casualties and displaced peoples have been significantly higher than in Casamance (Deets 2016). Furthermore, though there have been examples of brutality in Senegal there has not been the same systematic killing of civilians or the employment of child soldiers, both of which grasp the attention of international media reporting on Africa. Also factoring into our lack of knowledge on Casamance is the limited multilateral intervention seen in the region compared to other states. In Liberia both the ECOWAS and the United Nations have enacted political and military efforts to maintain their peacekeeping efforts. In fact, the UN's peacekeeping in Sierra Leone is the largest in the world. In contrast, the United Nations has limited its involvement in Ziguinchor and the surrounding statements to periodic press releases that encourage "dialogue" with the MFDC. Likewise, France—as the former colonist—has been relatively silent on this issue. In contrast, the MFDC has pursued both French and UN mediation and is angling to make their struggle known to the world. However, their very geographic isolation in this core-periphery relationship has hindered their success.

According to the Humanitarian Practice Network (HPN), the world's relative ignorance of the Casamance conflict is also a function of problems associated with information dissemination. To be clear, Senegal has a protracted history of a free press and strong academic institutions with linkages to the outside world. Casamance is also a relatively accessible region for visitors, as I noted during my three years of work in the region. Yet reporters within Casamance risk arrest or legal repercussions for reporting truthfully on the rebel crisis. And while Amnesty International has worked to report on the conflict and provide greater visibility, most coverage is in French, given Senegal's Francophone status, which keeps the crisis "out of sight" to those in the non-French speaking world. Even French tourists do not frequent the

Casamance region but choose to travel to beachside destinations such as Mbour or Saly. Once again, politics and cartography continue to alter the movement across the rural South.

There has also been a relative lack of donor commitment to the Casamance conflict worldwide. When the MFDC began planting landmines in Casamance beginning in 1997, many donors withdrew from the region, including such renowned groups such as the FAO, African Development Fund, and most critically, USAID (Linares 1992). When these groups left, many projects ended, and this had a profound impact especially upon agricultural resources and development, which is one of Casamance's greatest natural resources. Some donors have begun to return given the improved security and peace-making. USAID began a 10 million program in support of peace, and the World Bank, the European Union, and UNHCR are also assisting financially in peace processes or just as neutral observers (Marut 2002). These are positive changes indeed, but the marginalization of Casamance over time as a result of the war and more importantly, the root needs of local citizens, has had long-standing impacts on society and space (Mahling 2011).

SUMMARY

This chapter has explored the colonial cartography that led to economic divisions between northern and southern Senegal and the ways that these boundaries laid the foundation for a twentieth-century *Joola* ferry line. That southern region known as Casamance is physically separated from the major centers of power in Senegal (and Dakar) is central and cannot be overstated. The economic marginalization of Casamance is both a geographical and economic process. In Dakar lies the country's intelligentsia, headquarters for foreign donors, policy makers, and rule makers. With this history in mind there is need for a more integrated approach in West Africa that looks upon Casamance not as a discreet problem but a territory that is part of a much larger regional site linked with West Africa. The issue of illegal logging in the Bayotte Forest illustrates how Casamance's resource problems do not occur in a geographical vacuum south of The Gambia. Logging in Casamance is bound up in larger geopolitical processes that have been set in stone since colonial times. It is also related to the role of Casamance's neighbors such as Guinea-Bissau who have supported the MFDC movement given their related Jola history and culture. Logging in Bayotte is a force that is driven not by local actors alone but by global scale demand for endangered woods and any discussion of Casamance's issues—and the events leading up to the *Joola* crisis—must take into account these larger spatial processes occurring at multiple scales.

I have made the accidental case here that many of the MFDC's policies and practices served to reinforce Casamance's spatial isolation. The very rebel movement that longs to increase Casamance's power has in some ways led to its further disenfranchisement. It is still difficult to safely move across parts of this southern region, whether as an insider and outsider. Yet Casamance's frustration is also understandable, and their actions justified by an isolation which is rooted in the history of colonial era cartography and the state's failure to take proper account of the region's needs. In the case of the Bayotte Forest murders the media has served to reinforce misperceptions of a region rooted in religious rather than resource wars, or it has ignored Casamance in news stories altogether. I have attempted to explain how European administrators and cartographers set in motion a complex and absurd geography that would go on to shape movements of everyday life for Senegalese people. The chapter has illustrated how the man in the middle, President Jammeh, further exacerbated the movement of Senegambian people through both taxes and the prospect of arrest—should one travel through this Gambian semi-enclave. It was within this complex setting that the *Joola* crisis occurred, and the shipwreck disaster cannot be separated from this early colonial history, Senegal's relations with its neighbors, or the separatist movement that had long been evolving in the south. Jean-Paul Diatta, a twenty-four-year old Casamance Senegalese described the *Joola* as a "symbol of continuity" in the midst of these challenges. He said, "we still have solidarity, for the most part, but we are now sharing grief." Jean Diatta would lose seven members of his extended family on the ship that night.

NOTES

1. "Lu metti yàggul te ku muñ muuñ."
2. Radio for Peacebuilding Africa (RFPA) Radio for Peacebuilding Africa (RFPA) was founded in 2003 by the international non-profit Search for Common Ground. RFPA works on the premise that radio is the most accessible and democratic form of mass communication in Africa.
3. Cowries (*cypraeidae*) were shells used for several centuries by Africans, including West Africa, and date back to the eighth century. They feature in Hausa proverbs, were worn as amulets, used for communication and medicinal purposes, currency, and can be found in modern-day jewelry.
4. Anonymous interview with Senegalese man, August 2018.
5. Interview with Gambian man, October 2017.
6. Anonymous interview with colleague in Colorado, November 2019.
7. Serving as a regional economic union of fifteen countries in West Africa, these countries together comprise an area of over 5.1 million square kilometers and an estimated population of 349 million, roughly the size of the United States (Deets 2016).
8. New Zealand Government, Safe Travel Advice. https://www.safetravel.govt.nz/senegal

Chapter 2

Natural Resource and Population Geographies

It is better to farm your field than to pray for it.[1]

Wolof proverb

The African nation in which the *Joola* disaster unfolded is a study in geographical contrasts. In satellite images the landscape disparities between the north and south of Senegal are quite striking. Seen from above the Gambia River appears to slice the nation in half, stretching its sinewy length some 1,100 km (684 miles) from the mouth of the Atlantic Ocean at Denton Bridge to its source in the Fouta Djallon highlands of Guinea. North of The Gambia lies the arid regions of Dakar and Saint Louis, which are subject to dust storms or *haboobs* that sweep across the dry and sandy Sahel. While smaller in size than the north, southern Senegal is home to the majority of the state's agricultural activity, replete with lush, verdant territories that expand their way southward to the border with Guinea-Bissau (Figure 2.1). It is difficult to imagine a maritime horror unfolding in a place as remarkably beautiful as Casamance, but that is what occurred unexpectedly in September 2002.

In the Casamance natural region, the climate is low-lying and hot, peppered with mangrove forests and estuaries, known commonly as the tropical savanna, and not unlike the climate system found in south Florida. While the *Joola* affected *all* Senegalese citizens in one way or another, the majority of its passengers and victims hailed from this resource-rich, agricultural space (Becker 2012). This chapter outlines the geographical distinctions—environmental, demographic, and educational—between what is referred to as the urban north and rural south. At the same time, I focus squarely on agricultural economies as a unifying force for *all of* Senegal. By laying out these historical resource geographies, we can better understand the forces that set

Figure 2.1 Kapok Tree in Casamance, Senegal. Source: Author.

the stage for the *Joola* ferry line's creation as well as the composition of its passengers and their professions. Who were the people who chose to board the *Joola* ship? What agricultural practices were part of their everyday lives, and how did their vocations factor into their decision to book ferry passage to the urban north? What was the landscape of education in Senegal at that time, and how did it disproportionately affect residents south of The Gambia? What were the colonial mechanisms that have defined agricultural systems across West Africa today?

SENEGAL'S NATURAL RESOURCE BASE

Senegal has been broadly described as a predominantly rural state with limited natural resources, though in recent decades it has been christened one of Africa's new and emerging economies (World Bank 2020). Today the country gains most of its foreign exchange from phosphate mining, construction activities, tourism, agriculture, and commercial fishing (Capet et al. 2017). Senegalese industries also manufacture a broad range of commodities including textiles, wood, fertilizer, and in recent years, the extraction of gold products and the development of offshore oil. But by and large, it is *food* that ranks as the most important economic contributor, accounting for roughly 43.1 percent of all industrial output in Senegal (UNDP 2018). And while the nation is increasingly vulnerable to fluctuations in rainfall and changes in world commodity prices that affect production, food resources remain the mainstay of the economy and a marker of cultural identity. According to the World Bank, in 2017, some 34 percent of Senegal's population was employed in agriculture (Figure 2.2), and at the time of the *Joola* disaster in 2002, some 47 percent (nearly half of the workforce) was agriculturally engaged (UNDP 2002). This statistic is central to the story, because it allows us to understand *exactly who* traveled on the *Joola* and *under what circumstances*, in the same way we have come know the identities and economic motivations of those aboard the *Titanic.*

Agricultural maps of Senegal reveal the extent to which food and livestock production cover the expanse of the country as well as its spatial variation. Groundnut (peanut) production is dominant *both* north and south of The Gambia—a product of extensive colonial cash-crop systems—whereas rice paddy cultivation is relegated to the more humid climes of Casamance.[2] Cotton harvests are most dominant in the perimeter of Tambacounda, and livestock ranching reigns in the arid northern regions to the east of Saint Louis and Dakar. Given the maritime climate, fresh vegetables and fruits are grown from Dakar north to the Mauritania border, with other pockets of production found in the communities of Ziguinchor and Bignona. The rise of drip irrigation allows for small clusters of sugarcane production across Senegal, bypassing the heavy clay content of arid soils in sites such as

Figure 2.2 Livestock Being Prepared for Tabaski. Source: Author.

Diourbel, Kaolack, and Louga. Such agricultural diversity reflects Senegal's ability to adjust to the region's geographical limitations, reflecting a longstanding culture of adaptation and resilience, despite the troubling environmental legacy of French colonial rule (Fare et al. 2017). While maps suggest that farming operations are ubiquitous across Senegal, Casamance lays claim to most farming activities, both at the time of the ferry wreck and onward since. Nearly every place I visited in the south, farms were present, and while traveling on the ferry I marveled at the female merchants from Ziguinchor laden with cases of ripe mangoes or sacks of rice to sell in the far north (Figure 2.3).

That Africa in general—and *Senegal* in particular—is a site of enormous and diverse agriculture potential often surprises those unfamiliar with the lay of the land on this vast continent. I was astonished at just how ubiquitous fresh produce was during my travels south of the Sahara; prepackaged snack items are cost-prohibitive for most locals, but bananas, mangos, pineapples, oranges, and the fruit of the baobab tree are sold at inexpensive fruit stands all across Casamance. Vendors also sell plastic bags of peanuts onboard ferries in the south and in the crowded streets of Dakar. Research confirms this agricultural abundance and the potential for growth; Africa lays claim to an overwhelming 60 percent of the world's uncultivated arable land, and in 2011 the World Bank estimated 600 million hectares of suitable land that might be

Figure 2.3 Fruit Market in Senegal. Source: Author.

used for crops (Obasanjo 2012). The FAO possesses GIS data illustrating the immense range in agricultural potential, particularly south of the Sahara. This represents almost half of the world's total and more than the entire cultivated area of the United States. Agricultural potential is enormous, not only for providing sustenance to communities *within Africa* but also for feeding a global population of 7.5 billion (Plazier 2016). In that sense, Africa is the future breadbasket of the world, and as such groups like the International Food Policy Research Institute have turned their lens on Senegal and other states in order to better understand on-the-ground challenges and opportunities (UNDP 2003). A short walk through the bustling port of Ziguinchor undermines any archaic perceptions of an infertile land; Casamance is a proverbial place of plenty, where one witnesses not an absence but a plethora of fruit, vegetables, and fish loaded onto *pirogues,* vessels, horse-drawn carts, or anything that moves really. The *Joola* boat was as vital to the transport of produce and protein as it was to the movement of peoples in this critical regional trade network.

Why, then, does such an inaccurate image of Senegal and the rest of the continent's agricultural potential exist? Images of starving children—made famous by Ethiopia news coverage from the 1980s—have certainly eclipsed the on-the-ground realities of West Africa, perpetuating notions of a homogenous famine and food-insecure region, a typecast that does a disservice to the continent's agricultural potential. The 1980s was part of an era in which the media presented a landscape marred by drought and famine, made real by images of children dying from *kwashiorkor* in Addis Ababa (Nyssen et al. 2014). Over $100 million was raised for the Ethiopian cause, mostly through the power of music and celebrity, and those funds eventually found their way into the hands of politicians (World Resources Institute 2015). In Ethiopia this laid the foundation for a civil war—the conflict of which was not covered in full by that same media, and the consequence of which was further physical damage to drought-impacted agricultural land.

The Live Aid concert, organized by Sir Bob Geldof to mobilize funding for famine relief was a well-intentioned effort, working to pull Ethiopia out of poverty and increase food security. But while vast funds were raised, the process also served to revive extant prejudices and stereotypes about Africa, portraying Ethiopia and the continent and its people as helpless, unable to mobilize their own agricultural resources to save themselves (Badawi 2014). The Live Aid's marquee song—"Do They Know It's Christmas?"—also reinforced perceptions of human poverty while undermining the landscapes of Africa and their climatic diversity (Barton 2018).

And there won't be snow in Africa this Christmastime,
The greatest gift this year they'll get is life.[3]

The lyrics themselves reflect an insensitivity to the micro-geographies of the continent and a gross generalization about the state of its people.[4] On a more general level, the phrasing reflects an overall lack of awareness about the continent and its abundance, a perception that carries into contemporary times. The Band Aid lyrics, still played on American radio stations at Christmas, include several inaccurate excerpts:

Where nothing ever grows, no rain or rivers flow,
Do they know its Christmas time at all?[5]

As a result some outsiders do not visualize containers of cashews on board West African ferries bound for Dakar but instead see post-apocalyptic agricultural landscapes as portrayed in the news. Bob Geldof would later write that this song—along with "We are the World"—was one of his greatest regrets. While the impacts Live Aid had on the politics and skirmishes in Ethiopia have been widely discussed in academic circles (World Resources Institute 2015), there has been scant mention of how long-standing perceptions of Africa's agricultural lands were constructed through such songs. Iconic songs such as these clearly shaped American views of Africa, putting in place an image of a barren landscape devoid of potential. In 2018 I found myself walking the decks of the *Aline Sitoe* ferry bound for Dakar, joking with mango merchants and rice farmers who constitute the cornucopia that is Casamance. I grew up listening to Live Aid's music, but for me this visceral experience onboard the *Aline Sitoe Diatta* quickly quashed any outdated notions of an agriculturally impoverished space.

When former president Obama traveled to Africa in 2015 on the eve of the thirtieth anniversary of the Live Aid concert, his views had likely been forged by this flawed characterization of the continent. Instead the truth was that in recent years countries had taken measures to restore degraded lands, improve water access, and enhance food security (Plazier 2016). Recent research shows that northern Ethiopia, for example, is actually now greener than it has been for the previous 145 years (Nyssen 2015). And while pockets of degraded land still exist, many of our widespread perceptions of the continent do not reflect the levels of productivity and ingenuity that exist on the ground (World Resources Institute 2015). Some media programs like Journeyman's "Ethiopia Rising" are doing their best to underscore the resilience and adaptation of Ethiopian peoples, which are emblematic of strategies taking place across the continent (Dodd 2015). I outline this broad resource history here given just how critical farming and resources were to the *Joola* ferry line's creation.

The late Anthony Bourdain worked hard to counter the widespread stereotypes of this book's specific study site—Senegal—through the lens of his wildly popular program, *Parts Unknown,* broadcast on CNN. Bourdain never took the ferry or its replacement to Casamance, but he toured the restaurants, farms, and villages of Dakar and surrounding communities while reflecting upon the food and farms that comprise the rich cultures of West Africa. His effort to harness media serves not only to reinforce positive notions of Senegal, in particular, but also, in general, Africa—especially for those Americans who still maintain generalized views of a homogenous space fraught with poverty and hunger (Moseley 2001). One of the most notable excerpts from his narration includes:

> So let this episode in Senegal, an African nation which is over 90% Muslim, serve as both rebuke and example. It is a country that proudly elected as their first president after independence, a Christian—because they felt, in their best judgement, that regardless of his faith, he was the best person for the job. It is a country that defies stereotypes and expectations at every turn. Emerging from French colonial times as a functioning multi-cultural, multi-lingual, extraordinarily tolerant society. It has managed to avoid coups, tribal wars, dictatorships and most of the ills that afflicted so many of its neighbors and remains an absolutely enchanting place to visit, with delicious food, absolutely extraordinarily beautiful music, and a relatively free and easy attitude towards intermarriage, mixed race, intertribal relationships and foreign visitors. It has a powerful and proud tradition of hospitality that endures to this day. That they are predominantly Sufi Muslims, with attitudes towards behavior far removed from the more loony toon, extreme brand of Islam we see too all often on the news, is a distinction unlikely to be made by haters, most of whom have difficulty (or simply don't care) to even distinguish Muslims from Sikhs.[6]

We are in need of more stories related to agricultural abundance and cultural tolerance in order to replace those outdated images of the continent. Since food programs like *Parts Unknown* are popular among many sectors of society, this might also provide a new, promising venue in which to reframe agricultural identities of Africa. Bourdain's visit was a good first step, one that could be improved upon by also highlighting the Jola-based agricultural foundations particular to southern Casamance. Most of the onions and hot peppers that comprise Senegal's national dish of *ceebu jenn* actually traveled on the *Joola* ferry from Ziguinchor to make their way onto high-end restaurant tables in Dakar.

CASAMANCE YOUTH AND FARMING ATTITUDES

While agricultural stereotypes still remain, the reality is that for much of sub-Saharan Africa's history the continent has been short on *people* rather than arable, productive land. This is true for both East and West Africa. However, with populations growing and with some 60 percent of Africans under the age of twenty-four years, the challenge has been to negotiate ways to persuade young millennials in Senegal and elsewhere to follow careers in farming. Across the continent the average age of a smallholder (another term for farmer) is sixty years old. This is a sobering number given that across Africa the average *life expectancy* for men is just sixty-one, and for women, a few years higher at sixty-four (World Bank 2020). This is partly due to the fact that even with government support, farming still carries a stigma in some sectors such that teachers scold students that if "they don't study hard, they will end up growing cassava" (St Jacques 2009). Several of my colleagues in Casamance agreed with this notion, such as Mbaye, a thirty-two-year-old professor who argued:

> Here in Casamance we believe that farming is important. Many of those onboard the *Joola* were in some way related to farming, whether they were women working to sell mangos or wholesale fish or cassava or groundnuts. It's interesting, because we like our identity as farmers, but many of the young people do not want to farm. It carries with it a certain image that at least some young people are trying to shake.[7]

A twenty-one-year-old Jola woman who worked at Kadiandoumagne Hotel as a housekeeper drew the same conclusion regarding attitudes toward farming:

> It is no exaggeration to say that all of my relatives were once farmers. They pride themselves in being—I guess you could say—humble farming people from the south—but the reality is that they are so glad I did not choose that path. And they are glad I didn't marry a farmer either. Though the truth is my husband cannot find work. So which is better? To carry a stigma and have food on the table, or this existence.[8]

Thus while arable land is in large supply, and populations are growing, Senegalese are still grappling with notions about what counts as a respectable job in West African society. Farming is still not seen by some sectors as the first choice of professions (Poncian 2015). When I rode the *Aline Sitoe Diatta* ferry in Senegal, most of the Casamance-based merchants and farmers were stationed in third class, or in some instances ensconced below deck to check on their produce, reproducing the same social stratification onboard the *Joola* back in 2002.

In certain segments of West Africa, the dominant perception of farming is slowly changing, a dynamic that has been facilitated by agriculture technology start-ups, local celebrities and musicians, and a new breed of "agripreneurs" (NYT 2019). For example, *Cocoa na chocolate* is a pop song targeting youth and the video is replete with children riding tractors (One Campaign 2014).[9] One of the continent's biggest musical collaborations, it serves as a sharp contrast to the Live Aid hit of the 1980s, which altogether ignored Africa's farming potential. Instead the song is a call to African youth to support farming ventures, and is meant to serve as a poignant message to youth in Senegal, Ethiopia, and elsewhere on the continent to "do agriculture, because it pays" (One Campaign 2014). Not only did musicians target African youth with the song, but the African Union as well, with a message that Senegal and other states require better agricultural policies in order to tackle youth employment, provide better support to smallholder farms, boost productivity; increase value chains, and lift millions of Africans out of poverty. A friend in Ziguinchor who lost two male siblings on the *Joola*, both of them rice farmers, made the following statement during our conversation:

> I've heard the song many times, and I like what the musicians are working to do for small farms. There is honestly no better way to get the message to young people like me than through music, which I can see on my mobile. I suspect such things can make a difference in our thinking about ourselves. But I have to be honest, that I'm more confident in the farmers, and less confident in the government to listen to these words. Together we can make this work, if we focus on our identity and making a positive spin but we also need the government's support. And down here in Casamance, we have lots of concerns about what the government is willing to do for us. We have not always had positive experiences.[10]

While it is difficult to measure the success of these public efforts the involvement of celebrities and musicians is bound to play a positive role in the reshaping of agricultural identity. As I'll illustrate in chapter 5, musicians like Hamdine Watt are also playing a crucial role in determining how to memorialize the *Joola* by writing justice-oriented songs about the disaster.

In addition to these trends, it is important to understand local peoples' perception of food systems not only as an everyday livelihood but as a marker of *cultural identity*. In 2018 I spoke with Major Alpha Diatta, a former military man from Ziguinchor and a good friend, and he introduced me to the agricultural geography of Senegal from a Casamance perspective. He pulled out a pen and a sheet of paper in order to hand draw a map of the country from his vantage as a lifelong resident of Ziguinchor. In behavioral geography, mental maps are known as a person's point-of-view perception of their area of interaction. Also known as cognitive maps, they are used to understand subjective qualities that individuals impose upon neighborhood, state, or country.[11] Used in various research capacities they allow us to understand how individuals and communities interpret the world around them. Diatta's perception of Senegal was particularly poignant, and it highlighted the importance of agriculture in terms of the south's own particular regional identity:

> If you look at a map of Senegal (*draws map*), Dakar is the eye and the nose, Saint Louis is the brain, and Casamance (Major Diatta pauses for effect) is the heart but also the stomach. Do you see it?[12]

While unable to fully translate the cultural meanings behind Diatta's cognitive map, the metaphor of agricultural activity as the stomach of Casamance was undeniable. Farming is made manifest across the physical landscape of the south, and is also evident in Casamance's self-perception as a region of farmers, fishers, and freedom fighters.

While many Senegalese elders like Major Diatta perceive themselves as a nation of farmers—not unlike the Jeffersonian ideal described earlier in this book—many young people view agriculture as the job of someone who lives modestly, toiling under the hot sun for limited financial gains and a possible life of poverty. Young people who *do* chose farming in spite of these perceptions offer important insights into the psychology of their decision making. In an interview with Ibrahima Ndiaye, a twenty-seven-year-old cashew farmer from Kaolack, he stated:

> My father is a farmer so I began working in agriculture when I was just a kid, raising mostly raising chickens. After I finished my studies in project management at the University of Dakar's School of Economic Science, I decided to return to agriculture out of love for the activity.[13]

Yet this is not the mainstream view among Senegalese youth (World Bank 2016). Many millennials believe that the path to success is through business school or a desk job, and that material success can be found once the fields are left behind. Even Ibrahima outlined the challenges in place that interfere with his ability to succeed in agriculture, citing how land access is a necessary first step.

> When I decided to launch my agricultural project of growing cashews, I asked my grandfather and father for some land. Access to land is a barrier to youth who are going into agriculture, so I am lucky. However, I still struggle with getting access to water and farming equipment. [14]

Ibrahima, like many others including Oumar Ba, a thirty-one-year-old cattle farmer from Namarel agree that negative perceptions play a central role, but also factoring in are the realities of much needed practical training. With so much of the focus in postsecondary education on business and marketing, there is also a need for vocational schools that address the practicalities of becoming agricultural producers.

FARMING, MIGRATION, AND THE JOOLA FERRY

The rural exodus common across agricultural spaces in Senegal is another complex phenomenon that must be explored from multiple vantage points. It can partly explain why young residents of Ziguinchor boarded the *Joola* in the first place to seek more profitable careers in the north not found in their own backyards. Negative perceptions of farming do not exist in a historical vacuum. Perhaps farming had lost its attraction because young people watched their parents and grandparents struggle to make ends meet in the rural south. They saw better options in sites beyond Casamance, where new businesses served as pull factors for a young, energized, and hungry population with few local options. Or they migrated because they could not get the training they required, either for vocational work or for new, more attractive business prospects. We turn to these issues now: (1) economic development in Senegal, in general, and in Casamance, in particular; (2) perceptions of poverty and resource wealth; (3) the colonial legacy of farming and its longstanding repercussions; and (4) the spatial or geographic movement of people from place to place, based upon each of these factors. I more broadly explain the foundations of agriculture—both its challenges and constraints—and follow up with a discussion of how people were on the move across Senegal at the time of the *Joola* crisis. In the process we situate the *Joola* ferry migration in the context of the larger migration taking place across the state and the continent (Wanyoni et al. 2016).

Subsistence Affluence in Senegal

Economic development indicators place Senegal in the lowest margins in terms of worldwide rankings yet most Senegalese do not see themselves as *resource* poor. Nearly everyone I spoke with in Casamance was struggling to care for their families, but they tended to measure wealth using metrics beyond standard economic indicators alone. According to the United Nations Development Program (UNDP) the state's GDP per capita in 2017 was $2,384 USD, and it ranked 164 out of 189 in terms of human development that same year (UNDP 2019). More than one-third of the population lives below the poverty line and 75 percent of families suffer from chronic poverty. Yet these economic and human development statistics are not evenly distributed, given that disparities exist between rural regions and the capital of Dakar. In rural regions such as Casamance, some 66 percent of residents are considered poor, in contrast to 23 percent of those citizens living in Dakar (World Bank 2019). This rural cohort constitutes the bulk of passengers who boarded the *Joola* back in 2002 as well as those who travel the *Aline Sitoe Diatta* ferry today.

Despite these facts, in interviews with local people beyond Dakar—whether coastal Ngaparou or satellite communities in Ziguinchor—it is apparent that agricultural resources (and fishing—which we address in chapter 3) shape local notions of affluence. While residents are conscious of the economic challenges they face, there is also a collective sense of resource abundance. This perception remains steady despite the reality of global threats to food security—both in terms of agricultural production and artisanal fishing activities. As one university professor, born in Ziguinchor, explained:

> We were poor, but we did not think of ourselves as poor. If we were hungry, we would just reach up and grab mangoes from the trees, or fruit from the baobab, or go to the sea and catch fish. There are different kinds of poverty but we did not feel poor. We always had plenty to eat.[15]

Another Senegalese man detailed to me how plentiful resources were during September's fishing harvest in Casamance. The idea of superabundance, of unlimited food and fish, prevailed across all interviews—even though locals were sharply aware of the threats they now face due to global resource incursions into their territories:

> When the rains came last September, fish were literally jumping into our buckets. The rains were so fierce that there were fish flopping about in downtown Ziguinchor that day. Here, everybody eats. Nobody starves.[16]

To further underscore this point, since returning from my *Joola* fieldwork, I continue to receive texts from Ziguinchor residents who want me to witness the copious supply of agricultural produce and fishing resources. Some of their images reveal fishermen walking buckets of capitaine and thiof from their *pirogue* to the shore, lined up like soldiers in an afternoon drill. Food is clearly part of the region's collective pride, and Senegalese wished for me to know the extent of their wealth, in the same way an American friend might share their child's triumphs at school.

Casamance's perception of its own resource wealth, though not precise, hints at the notion that there are other types of extant wealth not measured by economic indicators alone. Often called *subsistence affluence*, this type of prosperity refers to the ability of a group to live a safe and healthy life with little formal income. This concept was derived from the small island nations of the South Pacific where strong support for the extended family or community is said to have contributed to alleviating extreme poverty (Bayliss-Smith and Feacham 1977; Lam 1982). The theory contends that while people are poor, they can find wealth in both their land and their kinship or community. And it was Hart (1972) who extended the subsistence affluence concept to West Africa through a field investigation into the ways in which informal economic activities led to greater autonomy for the marginalized Frafra migrants of Accra. The notion is that these off-the-books activities allow locals—including farmers and fishers—to bootstrap themselves, helping underdeveloped economies to flourish through their own indigenous enterprise (Hart 1972, 89). The concept underscores notions of resilience, but the downside is that subsistence affluence ignores the larger political structures within which communities must navigate everyday life. This same local bootstrapping occurred in the wake of the *Joola* disaster when the government failed to aid families, illustrating how important social capital is when financial constraints exist and state systems fail.

Though Casamance residents proudly *perceive* themselves as resource rich, the reality is that poverty and inequality is deepening in the country due to both economic and environmental conditions. Statistics show that there are Senegalese living in deep poverty despite traditional support systems, in which extended families help one another either through direct food supply or remittances from abroad. While poverty might not be visible given the "everyone eats" mantra, the fact is that much of Senegal lies in the drought-prone Sahel region which features irregular rainfall and generally poor soils. Only five percent of the land is irrigated, and therefore the country relies heavily upon rain-fed agriculture, which is subject to periodic, climate-induced shifts (World Food Programme 2011). Droughts are common, and the threat of pests such as locusts exists, which in 2004, destroyed agricultural crops for those already living on the margins. To complicate matters, Senegal is a net food importer, particularly for rice, which represents some 75 percent of cereal imports (World Bank 2016). Only recently has the country turned to rain-fed rice cultivation in order to decrease dependence on foreign suppliers; this is a particularly positive trend given that so much of Senegal's food culture is reliant upon rice consumption.[17] One can see the visual result of this shift in domestic rice production while on the *Aline Sitoe Diatta* ferry to Dakar. Sacks of rice are now loaded alongside groundnuts, and this self-sufficiency fits squarely with Casamance's desire for independence in all forms.

While these challenging climate regimes of the Sahel and the Savannah have been in place since the Mali Empire in West Africa, the political and agricultural economy has not. Casamance farming practices are also a byproduct of the colonial era, having evolved over time, creating both challenges and opportunities for these resource-dependent communities. To understand the current status of food production in Senegal, one must disentangle the early forces at work in shaping the region's resource economy and landscape (Kingston 2011). Prior to colonization West Africa was a region of small-scale farmers, fishers, and livestock producers, but the arrival of the French in colonial West Africa would transform the scale of food production in unprecedented ways—creating a dependency that until recently has not been rectified. I now turn to the colonial period to help understand farming, export commodities, and the culture of economic dependency in French West Africa. This is not a diversion but a discussion, one that helps to explain the economic and resource conditions that were in place at the time the *Joola* sank off the coast of the Gambia. This history also allows us to understand where Casamance's legendary resilience against outside oppression originated, and how it shaped local responses to Senegal's 2002 disaster. The injustices that followed in the wake of the *Joola* wreck were not Senegal's first rodeo; colonialism and its associated cash-crop systems had primed citizens for national disasters and illustrated how they could mobilize and respond to these incursions in characteristically fearless and forthright ways.

FRENCH WEST AFRICA AND THE CASH-CROP ECONOMY

Cash-crop production in French West Africa (FWA)—which includes Senegal—was the consequence of the end of the slave trade for the French Empire in 1848 and global economic transformations following the Industrial Revolution. The main cash-crops developed on the regional scale of FWA were groundnuts/peanuts, but palm oil and commodities such as gum, wax, and hides were also produced. Geographical innovations—such as steam shipping, railways, and telegraph—aided colonial expansion immensely and allowed export commodities to play a huge role in new economic structures (Curry-Machado 2013). In French West Africa, cash-crop development involved the *mass p*roduction of peanuts, which were then exported to Francophone cities such as Marseilles and Bordeaux to be made into soap and wax and for other industrial uses. West African peanuts from the nearby Gambia were in turn shipped in great quantities to American cities such as New York and Boston (Brooks 1975). One consequence was that French West Africa, like other colonial territories, became dependent upon external trade, a condition that would follow them into the independence period. The French system would soon alter a traditional subsistence economy in FWA that, while possessing its own unique disadvantages, was at least defined by local communities rather than colonial forces.

The peanut process in FWA actually began in the 1840s when French merchants from Bordeaux first initiated cultivation along the Senegal Valley and the Thies region, which was then extended to the Sine-Saloum region, The Gambia and Casamance. Local producers rapidly started to cultivate peanuts due to the interest of the European traders and the possibility of securing substantial profits (Thiao et al. 2018; World Food Programme 2011). The new colonial agrarian structure combined traditional features of "community" with capitalist elements based on the market economy. The original agricultural unit was kinship[18] with all family members serving as active participants in the production process. Each rural Senegalese family was given some land to cultivate that was classed as community property (Blundo 2001). In this arrangement, African peasants had the mandate to conserve and improve the lands for the benefit of the whole community. Agricultural production in Senegal had not only economic but social and spiritual significance—leading to the creation of myths about "agricultural heroes" related to natural spirits (Fall 2011). Farmers I met often placed *jinn* on crops and told me stories about their hardscrabble farming families dating back prior to the colonial period. Such strong work ethics were reinforced by the early teachings of Sufi Saint and religious leader Ahmadou Bamba (1853–1927), who founded the Mouride Brotherhood. One of the oft-quoted teachings of Bamba—told to me on pirogues, beaches, and at fruit stands—was "Pray as if you will die tomorrow and work as if you will live forever."

The second type of colonial agricultural production space or unit was that of Muslim farming communities known as "dara." These involved labor contributions of former slaves and young African men (talibes or disciples) who served the marabout or cheik, the religious and social leader of the community (Brooks 1977; Fare et al. 2017). These arrangements involved strict social regimes, whereby the dara produced peanuts as an attempt to preserve social status. Because this type of farming structure reinforced indigenous hierarchies in French West Africa, French colonial powers did not support the efforts of these Muslim farming communities. Therefore indigenously driven cash-crop structures would lead to struggles between local marabouts and foreign colonial actors (Fare et al. 1977; FAO 2015). Anti-French sentiment exists in many spaces across Senegal today as a response to former colonial rule. Some Senegalese—particularly merchants I met on ferries or in *pirogues*—will not speak the language despite being fluent in French. While traveling on the *Aline Sitoe Diatta* I was frequently told that "French is a colonial language" and in Casamance female merchants would praise me (and my son) for speaking Wolof instead. There was a bias against the French not only due to the legacy of political rule but also because of the way in which their ideologies had hijacked local language and indigenous cultivation systems.

For the French the solution to the "dara problem" came in the form of new colonial military campaigns to defeat these politico-religious chiefs in sites such as Upper Senegal and the Sine-Saloum. Control of indigenous elites soon allowed colonial forces to commandeer the region and enabled European traders to manage inland peanut markets directly (Barry and Harding 1992). The problem, however, was that this political-economic transformation led to a deterioration of living standards for the majority of West Africans, particularly for those residing in small rural villages who were now food-dependent on cash-crop structures (Lakroum 1982). Farming spaces that before had been devoted to an array of crops—using intercropping, artisanal method—were now dominated by speculative cash mono-crop systems. And, in the process, Casamance farmers began to trade commodities like peanuts, gum, timber, wax, and hides for cheaper products from urban areas in Dakar, including alcohol, textiles, and food. This externally imposed trade structure was unequal, leading to dependence and in turn solidifying imperial domination of the region. Many peanut producers fell deeper into poverty, laying the early foundations for the modern-day rural-urban divide. There is no doubt that the poverty witnessed in Ziguinchor's downtown district—where the *Joola* once disembarked—is a direct byproduct of these early colonial agricultural systems.

The onset of World War I led to the first crisis for existing colonial systems in French West Africa, which would have financial repercussions for the peanut producers themselves. Due to the lack of cargo space on ships needed for the war effort, along with the difficulties of navigating the seas in an era of conflict, commodity exports in both Dakar and Rufisque were interrupted. This shift had an impact on the prices paid to small-scale African producers, whereas in contrast, colonial companies found financial workarounds to maintain their profits. Export companies signed contracts with colonial administrators in French West Africa, allowing them to procure low-cost peanuts and other commodities, which would be transported to Europe on armed vessels (Bonin 1987; Coquery-Vidrovitch 1975). It was a win-win for the French colonizers and businessmen. And once again, colonial enterprises prevailed while West Africans were increasingly marginalized economically and politically. While not all social conditions were rosy under Mansa Musa's Mali Empire in the 1300s, West Africa certainly possessed enormous wealth and sophisticated centers for learning relative to French rule.

In the 1920s, European powers would turn to French West Africa and its economic resources in order to supply the postwar recovery effort. France, under the direction of the minister of the colonies (Alert Sarraut), drafted an economic plan that involved modernizing transportation in the West African colonies. The goal was "effective economic exploitation" of French West Africa, and in Senegal, this involved modernizing the port of Dakar, spearheading an effort to make this capital city the main—and only—hub for commodities. As a consequence, the role of smaller historic ports such as Saint Louis, Goree, and Rufisque (Castillo Hidalgo 2012) would be minimized, and the central transportation point would now be Dakar. Adding to this centralization were new railway connections to the port from both Mali and Niger—in which peanut production was encouraged and developed. By the end of the 1930s, peanuts were now traveling from all parts of French West Africa to this export hub—whether by wagon or international railway (Godard and Teurnier 1992). France had not only altered agricultural use patterns across the land but, in the process, set in motion a new preferred transportation network downgrading other historic hubs like Saint Louis. In the words of one resource economics professor, who sat talking with me in the urban center of Dakar:

> I think we're all pretty aware of the role of groundnuts. But we don't think about the geographical implications of the colonial cash-crop system in the same way that people buying cotton products is not on the minds of ancestors of cotton farmers in the USA. We are aware of the history, but don't necessarily tie it to the crop, if that makes sense. In a more general sense, colonialism stays with us as a legacy, and the groundnut manifests that history.[19]

In numerous discussions with Senegalese youth, they too expressed concerns about French colonialism but never tied it to the cash-crop system. However, it is clear that the agricultural foundations were transformed by these French economic structures well into the twentieth century.

When in the 1930s the world witnessed the second great global economic crisis, cash-crop prices for peanuts along with other commodities fell quickly in French West Africa. Rural communities experienced a kind of proletarianization in which they now served colonial powers on farms instead of working for themselves. As a result, rural poverty, dependence on foreign trade, and externally imposed transport networks resulted—and these phenomena extended far beyond the period of independence from French colonial rule. Today, when one witnesses sacks of peanuts being loaded onto the *Aline Sitoe Diatta* from Ziguinchor to Dakar, it reflects a complex political ecological history, one that was set in place 200 years ago. This discussion helps us understand Senegal's identity as an agricultural nation and also to see how the economy was shaped by outside forces beyond their control, leading to underdevelopment. Senegalese still love their groundnuts—on the streets, in restaurants, on ferries, after church or mosque—but the origin story of peanuts in Senegal clearly has complex, colonial roots.

Ecology of the Groundnut

Historically, peanuts were at one time to Senegal what coffee is to Brazil—though they still play a central role across society and space. In 1964 the *New York Times* reported that the "Peanut Harvest Rises in Senegal," citing a record breaking 110,000-ton crop, which at the time accounted for 82 percent of the country's exports (NYT 1964). By the 1960s, however, peanuts represented the only cash income source for the majority of farm families. In other words, they not only were the mainstay of the economy but were the economy itself. Though peanuts had first traveled to French West Africa by way of Portuguese slave traders from Latin America in the early nineteenth century, by the century's end, they could be found virtually everywhere in Senegal (NYT 1964). This was especially true in the maritime ports of Dakar, once capital of French West Africa, as locals prepared these nuts for export.

Peanuts are not really nuts at all, but hail from the legume family. They are called groundnuts because they grow in the earth at the roots of their leafy plants, in contrast to tree nuts cashews which can be harvested from trees across the expanse of Casamance. While the term "groundnut" technically also encompasses the Bambara and Hausa nuts, peanuts are by far the more important nut, which is even true in West African countries where other types of groundnuts originated, being endemic to the region. In Mali, Niger, Burkina Faso, and Cameroon—all West African states—the native Bambara groundnut *(vigna subterranea)* is still cultivated, given its high protein content, low impact on soils, and low risk of harvest failure even in variable rainfall regimes. Though these nuts have a dark history given the French imprint on agriculture, their importance in contemporary Senegal is most evident.

Today, groundnuts are eaten as snacks in the streets across West Africa and laid out as appetizers in restaurants or in a show of hospitality for guests. They are undeniably a staple of Senegalese society though their importance has changed over time. They were sold on board the *Joola,* and now, the new *Aline Sitoe Diatta,* and by merchants in the ports of Senegal. Fruit and vegetable vendors from Somone to Ziguinchor would give groundnuts to my family as a show of gratitude for our ongoing support of their business enterprise. In a show of reciprocity bartenders or waiters would hand my son a small bag of peanuts, a token of friendship and goodwill for the *toubabs*[20] from America. Peanuts are as ubiquitous as mangrove seeds in Senegal, and just as appreciated given their strong protein content. They sustain life when fish protein is in short supply, so it is often easy to forget the way in which these nuts were hijacked by colonial powers to create economic dependency and unravel existing transportation networks. Watching peanut sacks as they are loaded onto trucks and ships in Ziguinchor, it is important to remember the ways in which Senegal's food culture was shaped even prior to independence.

And today, peanuts are still the engine of the rural economy; their production accounts for 40 percent of cultivated land, or over two million hectares (Dinar et al. 2008).

POPULATION GROWTH AND DISTRIBUTION

In addition to the legacy of colonial economic systems, Senegal must contend with population growth and disproportionately high densities in urban areas. This in turn places pressure on the nation's food security given the paucity of people living in rural areas left to tend to farms. These issues are both important in their own right, and while interrelated, must be addressed separately in order to best understand the challenges facing those in the rural south, urban north, and those living somewhere in between. Unraveling this history will also help readers understand the why of where, or why residents of Casamance needed to board the *Joola* ferry to make their way north, where jobs, education, and prosperity awaited in the wake of uneven development in rural areas.

According to Open Data Africa, in 2013 Senegal was a nation of 13.5 million in a territory of 197,000 square kilometers, roughly the size of Nebraska (which, in contrast, the latter of which contains 1.92 million citizens). About 6.1 million of those individuals are concentrated in urban areas like Dakar, with the remaining 7.4 million are found in rural regions from the north to the south (UNDP 2018). Senegal maintains an urbanization rate of 43 percent, just slightly ahead of sub-Saharan Africa's average rate of 40 percent, but still low compared to industrialized states such as England (83 percent) or the United States (81 percent). To put these demographics in perspective, back in the 1960s Senegal maintained an urbanization rate of 23 percent with a population of 3.27 million (World Bank 2020). Children form an integral part of Senegalese demographics and as a result fertility rates—though they have dropped somewhat—still remain high in comparison to human growth indices in other emerging nations. Population growth places an undeniable burden on food security in the region, given the lack of infrastructure, variable rainfall, and colonial legacy of monocrops and cash-crop dependence. Everywhere I traveled in Senegal it was clear that parents love and revere their children and want what is best for their futures. Parents like Major Alpha Diatta (see chapter 3 and 4) had ambitious plans for his first-born son Abdoulaye, who was headed to medical school, but the *Joola* would cut his efforts short.

Despite Ziguinchor's status as a capital city of the south, it ranks only as the twelfth-most-populated municipality in all of Senegal, with the largest three cities located in the north. Like much of Senegal, the Casamance region lacks available jobs for the large and youthful population found across the entire country. Some 60 percent of the country's population is under the age of twenty-five, not uncommon for all of sub-Saharan Africa. Senegal's high fertility rate of 4.5 children per woman of childbearing age reinforces these bottom-heavy population pyramids. A quick glimpse at the passenger list for the *Joola* illustrates how southern farming families were even larger in size back in 2002. This means that in some cases, six or seven family members perished in the space of one fateful evening aboard the ferry. One problem that exists with this rapid-growth demographic structure is the state's inability to harness the potential human capital of so many youth, which leads to high unemployment, particularly in farming areas of the south where families are large. A consequence has been the migration of people from rural to urban regions in pursuit of work, which has repercussions on food systems in these areas (Lambert 2002; Wanyoni et al. 2016). In search of jobs in cities like Dakar or Thies, the result is a brain drain for communities left behind, as human capital is lost and family networks are disrupted (Sengupta 2003). Rural areas have experienced net losses as residents have migrated to cities, and the phenomena's consequences for farms and food security—as well as the communities that manage agriculture—is just now being explored. In the 1990s the establishment of the *Joola* ferry line provided one vehicle to help facilitate this rural exodus.

Migration

In many places across Senegal, resource-dependent communities have had to move not due to food security but as a consequence of externally produced factors beyond their control. In the former capital and now-UNESCO world heritage site of Saint Louis—fishermen have left the coast altogether due to new environmental constraints. In Saint Louis I spoke with thirty-one-year-old Mustapha about his work, and he explained how rising sea levels resulting from climate change are eroding shorelines, particularly in the area of Langue de Barbarie, a sandy peninsula that serves as a natural buffer with the ocean. As a result, the local school has crumbled, homes have been destroyed by high tides, and displaced people have been resettled by the authorities at inland camps (Breeden 2018). Here, fishermen like young Mustapha are left with the choice of whether to start farms in the interior or leave Senegal. Mustapha has long considered boarding a *pirogue* en route to the Canary Islands in search of a better livelihood to support his family. Aware of the dangers of human smuggling on the high seas, he is left with few choices in the face of climate-induced environmental change.

Spatial movement of Senegal's populations is not restricted to what demographers call *internal* migration alone. The *Joola* ferry in some ways was a microcosm for a spatial process that was (and is) unfolding across this entire world region. *External* migration has also become a critical issue throughout the nation, as young people seek out opportunities or education not found in their own backyards (Levitt and Schiller 2004; Linares 2003). Some Senegalese have even migrated beyond national borders where farming or fishing opportunities are seen as more favorable. For example, in certain cases residents moved to Guinea-Bissau during the height of the MFDC crisis in Casamance, seeking political refuge in a nearby country not altogether different from their own. In other cases, Senegalese have migrated to other parts of Africa in pursuit of jobs not available in their communities. I met thirty-two-year-old Ibrahima, in Ngaparou, who had spent four years away from his wife and son working as a chef in South Africa. He was filled with pride for his work abroad, given his former prestigious position at the embassy in Johannesburg. He shared with me photographs of his food creations and of the diplomats he'd met during his stay in the city. But he also said:

> The work abroad was good, but I missed my family. It is time that cannot be taken back. But I had no choice, there were few jobs in Dakar at the time for my skill level, so I went, and stayed. WhatsApp saved us, and allowed us to communicate, I could watch my son grow up.[21]

Like many Senegalese, Ibrahima has a strong sense of gratitude for what he does have, rather than what he does not. He currently works for Hotel Phenix as the head chef, a family owned hotel and restaurant complex in Somone, which sits on the beach not far from the bustle of Mbour and Saly. Ibrahima explained how his friends narrowly avoided boarding the *Joola* back in 2002, instead opting for jobs in Johannesburg, a twist of fate that formed a common thread in stories I heard.

One of the most controversial forms of migration is for those who travel beyond Africa, to Europe, a process known as international migration (Barton 2020). While the spatial path taken from Africa to Europe varies, in the case of Senegal, travelers usually opt for one of two routes—overland or by sea. From Dakar, some migrants will travel across the length of the Sahara desert, following the route of caravan traders until they reach northern coastal hubs. From there, migrants make the jump across the Mediterranean by boat, and oftentimes in a clandestine fashion. In the case of Senegal, the other, more common route is to leave via the coast on a *pirogue* bound for the Canary Islands and other Mediterranean ports. The challenges that *pirogue* migrants must endure on the journey to Europe were made famous in Senegalese director's Moussa Touré's film, *La Pirogue* (2017), which did its best to show the economic factors that propel (mostly) young male West Africans to leave their homes and face the perils of the high seas. If a Senegalese-based film on the *Joola* comes to fruition, there is no doubt that Mr. Touré will be at the director's helm of such an ambitious venture.

While the causes of international migration are diverse, this spatial movement of people has become a concern for policy makers in Europe, who fear that a large movement of undocumented workers will have repercussions for existing social services. However, the reality is that the majority of those who migrate beyond Senegalese and other African borders are not moving to Europe; rather, they are moving *within Africa* (Barton 2017, 2020). Yet the media has chosen to place a sharp focus on the movement of people across the Mediterranean to states like France, Turkey, and Spain. As many citizens of Africa are adherents to Islam, there has also been a marked concern that those headed north are armed and militant, which is not the reality for the vast majority of asylum seekers. The desperation of those migrants who do flee north are in pursuit of jobs not available in their communities. They are in search of remittances—rather than troublemaking—in order to send back home to extended families on the continent. In the case of Casamance, home to the vast majority of *Joola* casualities, many victims were in fact Muslims, known famously for their peaceful coexistence. I had countless conversations with young Sufis in Ziguinchor or Somone who had considered boarding a *pirogue* to Europe if things did not improve. And, with the outbreak of the novel coronavirus in 2020, financial pressures for many Senegalese were mounting, especially with international tourism revenues on the sharp decline.

A Reason to Stay: Educational Opportunities in the South

One of the processes that helps keep individuals in place—or, at least to temporarily curb the flow of migration elsewhere—is to give people reasons to stay. And one of the greatest draws in Senegal and elsewhere is simply that of education and access to opportunities (USAID 2018). Much like other regions in Senegal the people of Casamance had long been promised that a university would be built to help accommodate the needs of young people from as far west as Cap Skirring, eastward toward Kolda, where populations are even higher than Ziguinchor (Figure 2.4). Youth needed a reason to remain in their home communities while enhancing their work or career prospects. This was also true in sites like Bambey or Thies, where rapidly growing populations were left with no choice but to travel to Dakar in search of higher education. Beginning in the 1990s, these young people typically chose the *Joola* as their preferred form of transportation for the reasons outlined in the chapter entitled "Colonial Cartographies."

Most southerners agree that there has long been a high importance placed on education *within* Casamance. Since 1991, the net school enrollment ratio

Figure 2.4 Map of Casamance Region. Cartographer: Dr. Jieun Lee.

had risen from 80.4 percent to 96.9 percent in Ziguinchor and from 24.4 percent to 83.7 percent in nearby Kolda (World Bank 2019). Further, those who work on the ground in the region have argued that Casamance had managed to consistently attract some of the best K-12 teachers in the country (FAO 2015). It was therefore a natural next step to add a postsecondary institution to Ziguinchor's suite of educational options in order to place them on par with Saint Louis and Dakar. Casamance locals complained that Dakar possessed the bulk of the educational resources and, as one Senegalese man described it, "the capital had long been milking the resources of the rest of the state" (Barton 2018). The populations of the south felt marginalized at the time of the *Joola* wreck, and this was underscored by the region's educational deficiencies and reinforced by the ongoing MFDC rebellion in Casamance.

Part of the reason a university in Ziguinchor was absent for so long was because of decisions associated with the 1970s financial crisis (Park and Reid 2013). During that time, funding to higher education was cut in Senegal, and international agencies stepped in over the next decade to institute reforms. Yet most of the available funding went to meet the needs of primary schools, and later came the boom in Senegalese secondary education initiatives. This was certainly a positive step for many rural communities. The statistics were impressive and transformative; in 1984 just 50 percent of Senegalese children received a primary education but by 2004 more than 90 percent acquired one (World Bank 2020). This means that within two decades access to primary education had nearly doubled across the nation. These shifts were seen in both rural and urban areas, as the new developed policies were not just restricted to the more privileged capital region of Dakar and the surrounding metropolis. But in a bizarre twist, the funds encumbered to help primary educational institutions were used at the expense of developing a space for higher education in the south.

Because a college education was not available locally in the 1990s in the south, the creation of the *Joola* ferry route would enable Casamance youth to hop onboard a ship to reach Dakar or Saint Louis instead of taking the long drive by road. Gaston-Berger University would certainly take longer to reach, given its position in the northern city of Saint Louis, but some Casamance residents still enrolled there upon its founding in 1990. Naturally, then, the main option for postsecondary education was the University of Dakar, which had been already built by 1957. This university was accessible, having always been free to students who possessed the international baccalaureate degree. At the time of independence in the 1960s, the university had an enrollment of just 1,108 students, of which only 39 percent were Senegalese, part of the legacy to serve white and Métis[22] students as well as the small educated elite of four free towns of Senegal with nominal French citizenship (Jones 2013). University of Dakar soon became the most prestigious and largest university in French West Africa, with 60,000 students in attendance thanks in part to a US $32 million endowment. Attending this institution was the best option for young people in Ziguinchor or Toumbacounda who were in pursuit of new career opportunities and upward mobility. My Senegalese-American colleague, a Casamance native, was one of those students who boarded the *Joola* in the 1990s to pursue his degree in linguistics in Saint Louis.

Much of the credit for educational improvements in the 2000s can actually be attributed to President Abdoulaye Wade, whose overarching goals included infrastructural investments with a focus on educational access for all. Wade worked to expand the number of secondary schools in Senegalese communities (Ndiaye 2012). Upon completing this initiative, President Wade set in place a visionary plan to build three new universities in underserved regions across the country in order to address the grievance that Dakar was in a position of privilege. In 2007, Alioune Diop University was opened in the Bambey department of Diourbel. Thies University was established that same year to serve the outlying areas of the northern area, and finally, Ziguinchor University was christened in February to assist those in the south. While Wade was publicly criticized for many of his policies, his educational reforms had long-lasting impacts upon society and space. Not only did the president transform secondary school densities in all regions, but he created a massive system of higher education that included both public institutions and vocational schools, the latter of which allowed for specialized, professional training. Some of those professional schools managed to attract students from

North Africa who sought training not available in their own countries. These were certainly great strides for Senegal's higher education objectives, but for Casamance, it was seen as too little, too late.

When the University of Ziguinchor finally opened its doors seven years after the shipwreck disaster, some argued that it was the state's attempt to appease residents of Casamance in the wake of the *Joola* disaster. Many did not see it as an earnest or legitimate educational opportunity for youth in the south. Locals I spoke with, including members of the ANFV, were proud of their regional university, but nevertheless some, like thirty-six-year-old Marieme, argued:

> The timing of the University fell right in line with the post-disaster recovery. They had many years to build us a university but they waited until after the wreck. I can guarantee you that many of those people would not have boarded that ship had they been able to attend university here in Ziguinchor. Sure, some would've gone to Dakar to pursue specialized fields not available here, but you know like I know that there were hundreds of college students onboard the *Joola* in September. They were heading back after summer to university classes in the city. At least half of them wouldn't have been on the ship had there been more education and resources here.[23]

Many people I interviewed applied causation rather than correlation to the case of the *Joola's* "brain drain," in which educated Casamance citizens left a city that they perceived had failed them in terms of advanced education. They blamed the government repeatedly for the fact that large numbers of youth were on the ship, citing the state as the responsible party and for not giving them reasons to stay:

> I mean, if you look at the situation, my brother would never have taken the ship had there been sufficient schools here in Ziguinchor. Unlike in say Europe or the USA, Senegalese will not choose a far-off university since it might mean sacrificing connections to family and extended family. We like where we live, we have no problem getting our education here if that was an option back then.[24]

It is interesting is that since the University of Ziguinchor has been constructed, a large proportion of locals still choose to study elsewhere, whether in Dakar or Saint Louis, or even abroad in some cases. Therefore it is possible that many youth would have opted for an education beyond Casamance even if a university had been available in their backyard. Not unlike in the United States, the closest university might not provide the desired training or specific area of study for a potential undergraduate, so students seek their degree elsewhere. Yet interviewees still described general discontent with the state government or apparatus, including its approach toward a regional education. Locals had been repeatedly ignored, their resources underdeveloped, their people underserved, and the siting of the University of Ziguinchor was seen as a type of virtue signaling for those left behind.

Yet the government's perspective on Casamance's higher education initiatives differed markedly from the attitudes of local citizens in Ziguinchor cited above. The state argued that its goal was to "establish a regional university center (CUR) in Ziguinchor by public decree in order to meet the needs and potential of the south" (Public Decree 2008-537). The objective was to "promote the region as a destination and integrate its economic and intellectual needs into the national framework while opening up a gateway to Casamance" (University of Assane Seck Ziguinchor 2020). It is true that when the university first opened its doors, students could choose several traditional areas of study including mathematics, physics, chemistry, computer science, geography and GIS, management and applied informatics, a positive first step for a region not yet on the educational map of Senegal. But when I spoke with locals about the stated mission of the new university as well as its program offerings at that time, criticisms ensued. Abdoulaye, a twenty-two-year-old shopkeeper in downtown Ziguinchor, complained:

> In addition to the delay in building a university in Ziguinchor, there is also the issue that a lot of the subjects we wanted to study were just not available. My dad, who is Puular, who did not go to university, used to wonder how the state actually selected those subjects. I mean, Math and Computer Science are important and practical, but there were many of us who wanted to study liberal arts and agriculture, and just would have to go elsewhere. Once again, we were not consulted on what kind of university we wanted. They just built it and expected us to come.[25]

It is difficult to challenge the validity of Abdoulaye's argument given the ease of mobility in places like the United States. Many students in my home state of Colorado can select another public institution in which to enroll if they meet the requirements and the college offers the desired degree program. It is not the same in Casamance. Senegalese students do not always have the same level of spatial mobility, as they often have to care for aging parents or siblings, causing them to stay close home. A four-year absence from home, without income, would certainly have an impact upon the larger family structure which involves a great degree of economic interdependence. Thus while it is true that the state fulfilled its obligation by at long last establishing a regional university in Ziguinchor, it's easy to relate to the locals who resented the delay in the university's construction. They felt that university's construction was a symbolic band-aid applied only in the wake of the *Joola* disaster, and that the school's foundation and chosen subject matter served as an example of how the state had ignored their input yet again.

Critics might also argue that the first subjects made available at the University of Ziguinchor—math, science, GIS, computer science—were designed to prepare undergraduates for professional careers that could not be found in their hometown. In some ways these degrees prepare students for careers that would *force* them to migrate elsewhere since the best jobs were available elsewhere. Another way of unraveling this is to say that Casamance's existing curriculum encouraged a brain drain. Computer science skills learned at University of Ziguinchor could be utilized to start businesses in Casamance, or geography competencies such as GIS might be harnessed to map Casamance, but the chance of securing a job in those professions locally was slim. Some people I interviewed argued that receiving a degree from University of Ziguinchor would propel students to move beyond Casamance in search of opportunities, fracturing extended families, and sending only remittances to assist those left behind.

Despite these criticisms, it is inarguable that the University of Ziguinchor provided one option for bridging the intellectual divide between the north and south. While an imperfect system in its incipient stages, it was indeed the only educational system available, and a good first step for Casamance as well as the state apparatus. During its first year the University of Ziguinchor enrolled just 257 students, who were supervised by twenty professors and researchers and thirty administrative staff. In 2013, the University of Ziguinchor was later renamed to honor Assane Seck, Senegal's former Ministry of Culture and minister of education, a trained linguist and geographer who passed away at the age of ninety-three. Enrollment is growing, and faculty are now eligible for federal monies in order to enhance their research agendas and aid undergraduate students (Ministry of Finance 2001). While its establishment came too late after the *Joola* crisis and the south's overall disenfranchisement, it had helped put the south on the map (Gansler 2018).

Today, the university's curriculum speaks directly to the growing needs of a nation and a region, with new areas of study that focus on community-engaged scholarship that together address local food and energy needs. Assane Seck University/Ziguinchor University, in partnership with three other universities—Thies, Bambey, Saint Louis—offers a relatively new graduate Programme on Renewable Energies and Energy Efficiency in Senegal (PESEREE), which receives support from both Deutsche Gesellschaft für Internationale Zusammenarbeit (GIZ) and the Federal Ministry for Economic Cooperation and Development (BMZ). To date some 100 students have enrolled in the program. These undergraduates can switch between universities to complete coursework which provides greater flexibility. Thus these new programs illustrate the government's recognition that Senegalese *are on the move* and require flexibility to meet the demands of their families back home. Furthermore, the subject of energy and food puts a lens on existing problems at the local level, providing students with everyday skills in their backyard laboratory of Casamance. There are still no courses offered on human disasters including the *Joola* shipwreck, as the subject remains a highly contested political issue.

One of the most promising new programs at Assane Seck (University of Ziguinchor) fuses together the need for energy independence with existing but discarded biomass from all of the Casamance farm operations. This takes the form of cashew apples, a local staple found across West Africa and found as fruit of the acajou tree. While some of the fruit is processed, the remainder is usually left to rot, and researchers are now developing ways to harness fruit waste alongside cow dung to create biogas programs that can supply energy to off-the-grid villages. The local program feeds directly into Senegal's food culture and agricultural history. Local researchers like Omar Kata Faye have been enormously successful with these locally driven, natural resource-focused projects that have everyday relevance for their families and communities. His fieldwork provides him with the competencies that he'll need in the job market whether he remains in Ziguinchor or is employed elsewhere in Senegal (Gansler 2018). The point is that Omar now has options, not unlike his aunts and uncles who traveled the *Joola* back in the 1990s, their sights set on Dakar and its promise of upward mobility.

SUMMARY

This chapter has made the case that Senegal's rural south (Casamance) and its urban north (Dakar) are unique from one another both in terms of their natural resources, demographics, and educational capital. Yet, at the same time, I have shown how Senegal shares a *collective* national history of farming that parallels larger trends across the continent of Africa today. Challenges to food security certainly exist and are exacerbated by Senegal's growing population, but not to the degree made popular by media representations. This chapter has shown that Senegal is not short on arable land but rather a large cohort of youth who wish to continue the legacy of their family's farming traditions. These young millennials are often drawn by careers devoid of rural-based stigmas, which is part of the reason they boarded the *Joola* back in the years leading up to the 2002 tragedy.

This section has also detailed Senegal's cash-crop production of groundnuts, a colonial project that laid the foundations for the nation's economic dependency. I've shown how European imposed cash-crop systems undermined traditional agriculture and disrupted transportation in predatory ways.

Yet despite this checkered colonial past, farming remains an important part of Senegal's cultural identity and rural economy. Farmers from Casamance formed a large proportion of those passengers onboard the *Joola* ferry. Some of them were seeking to get their agriculture products to Dakar (mangos, rice, peanuts) where larger and more robust markets could be found. Other passengers were the children of farmers, pursuing a postsecondary education not available in their own backyard. The takeaway is that Senegal's north and south were linked politically, economically and culturally by a maritime ferry line designed to circumvent ongoing transportation and political challenges as well as uneven national development. *Joola* passengers could evade the MFDC rebel conflict, Jammeh's brutal reign by journeying north, but the ferry would prove to be a precarious stop-gap response with legendary repercussions.

NOTES

1. Yalla yalla bay sa tool.
2. Cash-crop systems and their development are discussed extensively in this book.
3. Band Aid, 1984. "Do They Know its Christmas?" Sarm West Studios, London.
4. Helligar, Jeremy. 2019. "The Unbearable Unwokeness of 'Do They Know It's Christmas?'" Medium, December 17, 2019.
5. Band Aid, 1984. "Do They Know its Christmas?" Sarm West Studios, London.
6. Osterholm, Erik. 2016. Parts Unknown: Senegal. Release Date: May 29, 2016.
7. Anonymous interview with college professor, July 2018.
8. Anonymous interview with female hotel worker, August 2018.
9. The hit song features musicians including A.Y., Bufallo Souljah, Dama Do Bling, Diamond, Dontom, Fally Ipupa, Femi Kuti, Judith Sephuma, Juliani, Kunle Ayo, Vusi Nova, Liz Ogumbo, Nancy G, Omawumi, Rachid Taha, Tiken Jah Fakoly, Wax Dey and Victoria Kimani.
10. Anonymous interview with male friend in Ziguinchor, August 2018.
11. See Lynch (1960), *Image of the City*, MIT Press.
12. Major Alpha Dia, in conversation with author, July 2018.
13. Anonymous interview with farmer, July 2018.
14. Anonymous interview with colleague, August 2017.
15. Anonymous interview with professor, Ziguinchor, August 2018.
16. Anonymous interview with male Senegalese, Ziguinchor, August 2019.
17. Ceebu jenn and other fish and rice dishes are prime examples of this.
18. By kinship unit we refer here to those production systems based upon family labor.
19. Anonymous interview with professor of economics, Dakar, November 2018.
20. Toubab or Toubabou is a Central and West African name for a person of European descent and is used most frequently in The Gambia, Senegal, Mali, and

Ivory Coast. The term can be derogatory, but it is also associated with "wealthy travelers."

21. Anonymous interview with restaurant worker in Somone, October 2018.

22. This term refers to an influential group of mixed-race people who settled in coastal Africa under French colonialism. As colonialism spread, the métis entered into politics and began to assert their position as local elites and power brokers against French rule. For a full discussion, see: Jones, Hillary. 2013. *The Metis of Senegal: Urban Life and Politics in French West Africa.* Bloomington: Indiana University Press.

23. Anonymous interview with a female produce vendor, November 2018.

24. Anonymous interview in Dakar, September 2018.

25. Anonymous interview with shopkeeper Ziguinchor, October 2018.

Chapter 3

Geographies of the Sea

The sea is never pregnant.

Wolof proverb.

In order to fully understand the *Joola* disaster it's important to contextualize the space in which the tragedy occurred: the Atlantic Ocean and its related riverine locales. Senegal is as much a nation of fishermen as it is a land of farmers as outlined in chapter 2. This is evidenced by the countless fishing *pirogues* that ply the country's coastal waters each day in pursuit of dorade, thiouf, and capitane, from Ziguinchor in the far south to Saint Louis in the north. West Africa maintains a set of rich and diverse fishing grounds that not only form the mainstay of local communities but contribute to the region's cultural identity. This chapter seeks to explore how these maritime spaces are utilized (or avoided) by Senegalese people and the threats they now face. In short, understanding the *Joola* disaster requires a deep investigation into Senegal's relationship with the sea and how it has changed over time.

I spent several months in three piscatorial (fishing dependent) communities in Senegal that have connections to the *Joola* story: (1) Somone, a coastal fishing village not far from Dakar, north of The Gambia; (2) Saint Louis, in the arid north; and (3) Ziguinchor, a riverine fishing community in the Région Naturelle de la Casamance. The sections to follow lay out the challenges facing these maritime communities, including but not limited to the shipwreck disaster. At the same time this chapter explores community resilience along Senegal's coasts, or the ways in which fishers have created workarounds to help sustain their livelihoods despite external threats. To accomplish this goal, I weave together personal accounts from time spent with fishermen, marabouts, and coastal residents. I employ interview and participant observation data alongside West African fishing statistics in order to best visualize the space of the sea before and after the *Joola* crisis.

During my time in Senegal (2016–2018) I arranged thirty-two field interviews with fishermen, fishing cooperatives, and community village leaders in Somone, Saint Louis, and Ziguinchor in order to understand oceanic-centered issues. The interviews took place with the translation assistance of local colleagues in each region who I've acknowledged in the preface to this book. The majority of fishers and leaders I spoke with—on beaches, boats, and in the water itself—described five key maritime themes: (1) the perils of water safety and swimming education for fishermen; (2) pollution of the maritime environment by multiple entities and the loss of fish, sharks and dolphins; (3) the incursion of sea space by Chinese pirate fishers or industrial vessels, the latter of which are occupying Senegal's exclusive economic zone having paid the government for fishing permits; (4) the reality of global climate change and its effects on coastal erosion and water salinity; and (5) Senegalese identity and mysticism as it relates to the sea and its aquatic spaces. Whereas the *Joola* shipwreck was at times a sensitive subject, the sea was a topic that locals discussed with great passion and ease. Oftentimes a conversation about a Senegalese *pirogue* would lead to discussion of the *Joola* disaster as one theme could not always be separated from the other. This chapter therefore integrates all five of these issues using field notes while highlighting their relevance to the 2002 wreck.

THE SPACE OF THE SEA: SWIMMING WITH SENEGALESE

It is October 2018. Just offshore the fishing village of Somone sits a series of buoyed flags used as markers for *pirogue* fishermen and their nets, some colored a menacing black, others white in surrender, each of them standing sentinel before fishermen return from the sea after periodic stretches away from home. The flags also signal unsafe locations where rocks lie beneath the surface, threatening to shred a boat's engine in minutes. They serve as geographical beacons marking the point at which the crew can return to shore. As a recreational swimmer, I have taken to freestyling parallel with the shore as part of my routine workouts, but I've been lured by the flags off in the distance. I have watched my friend Boubacar (Boucar for short) swim to the flag and back, a rare sight in this area. He, like other residents, is dedicated to his sport of fitness in this outdoor sandbox, a process that is ritualized on the beaches of Somone, Ngaparou, Dakar, most of which are gendered spaces belonging mostly to men.

Joining Boucar on a long swim out to the buoy is a lesson in humility, even for expats from a country where swimming lessons are the norm for young children. One day I summon the courage to accompany him, both of us adorned with goggles and swimming caps. We swim parallel in freestyle form, taking a direct path to the buoy, periodically glancing up to make sure we do not crash into one another as the current makes it hard to stay the course. As we move further from land, the waves loom large up close compared to the comfortable view from shore. Swimming freestyle in a pool is fairly straightforward, up and back for a mile and half in flatwater with reliable rehydration stops from a Nalgene bottle. Swimming in ocean water is much more challenging. Saltwater leaks between the crevices of our old goggles, and each breath is filled with a mouthful of seawater from the large rolling waves, as we pinwheel our arms into a ritualized crawl.

Sharks are on my mind. Anecdotally locals argue that there is little danger of shark attacks in West Africa, while other research shows that they still roam the area but most attacks go unreported, thus skewing statistics. However, Boucar assures me that he has not seen sharks (*requin i*n French) in years and smiles broadly at such a trivial concern. As we slice our way out toward the flag Boucar lifts his head from the sea to ask if I am tired, and the truth is I am more scared than fatigued. The abundance of trash in the inshore masquerades as famished, toothy sea creatures, causing momentary panic when my foot or hand intercepts a plastic grocery bag, of which there are many. The bags drift with a kind of slow motion jet propulsion, like lifeless jellyfish. But onward we swim toward our goal, the fisherman's buoyed black flag, a tiny welcoming island in this murky and unfamiliar sea. Always, while I swim here, the *Joola* ferry is fresh at the forefront of my mind. People not unlike me struggled for survival in these waters, waiting for a state-sponsored rescue that never came. As noted, 1,863 of them perished, leaving only sixty-four survivors, only one of them female (Becker 2012).

Boucar will later tell my family that he is an amateur swimmer, having swam the 5-kilometer distance from the capital port city of Dakar to Gorée Island, a site where African slaves were once traded and then boarded onto ships like sardines to their fateful journey on the Middle Passage en route to the Americas. He invites us to compete in the annual event of 500 or more swimmers, reserved for the last weekend in September 2018 (OWSA 2020). It is ironic that the timeframe of this competition will roughly overlap with September 26—the day set aside to remember and memorialize the *Joola* shipwreck, many of the victims from the boat having never learned to swim.

This athletic occupation of marine space in the form of a swimming competition is symbolic, given that in historic times slaves would attempt to swim *away* from Gorée island toward the shore to escape harm in what were once shark-laden waters. Today, competitors—both Senegalese and expats—in the 5-kilometer race swim *toward* Gorée, an island that is now accessible via a short ferry ride from the bustling port of Dakar. Tourists and locals come here for the day, given its proximity to Dakar and its enduring history. And on the beaches of Gorée, there is a feeling of languor, even lassitude, where school children play in the surf bedecked in swimsuits, undergarments, or simply the clothes they wore on the ferry ride across the bay. In some ways Gorée can be likened to Alcatraz Island offshore San Francisco, which once served as a prison for notorious convicts. Gorée is a beautiful oasis not far from the thriving cosmopolitan center of Dakar, and both Alcatraz and Gorée are two islands where life now moves more slowly, tourism dominates, and the views are spectacular. To imagine the historic view of mainland California or West Africa from this small distance to shore is quite sobering, given that both sets of residents were once prisoners of different forms. Gorée constitutes a space in which some of the most horrific activities took place, the trading of slaves (with assistance by African slave traders) from the "door of no return." The feel here is haunting but hopeful, particularly in light of swimmers like Boucar who are now reclaiming the sea.

We reach the white flag after twenty minutes of freestyle swim, and clutch the buoy for less than one minute, our only source of safety this far from shore. We are on top of a watery Mt. Everest and need to get back to land swiftly but without panic. I feel irresponsible being this far from land, but somehow this swim seems crucial for understanding the visceral fear of the Western Atlantic waters given what I came here to study. The *Joola* shipwreck. Boucar smiles the entire time, one of the strongest swimmers I've met to date. It is a "no worries" grin common to many Senegalese who will say "*begue*" before a farewell. It simply means, "happy." Swimming is a rare skill in parts of West Africa, certainly not so for those who dwell in the beach cultures of Saly, Somone, and Ngaparou. But more broadly only *12 percent of Senegalese* are able to swim (Senegal Scientific Ministry 2019). For the residents of Ziguinchor in the south, where the muck-laden, riverine environment of the mangroves dominate the tropical landscape of the rural south, swimming is an even less popular sport. One interviewee told me of his formative experiences growing up in Casamance, where he nearly drowned right by the Ziguinchor bridge, having never learned to swim as a child. Two women in fishing boats used a *pirogue* paddle to pull him to safety, and now in his forties, he still does not swim. In the aftermath of the *Joola* tragedy, there were some groups who criticized the ship's passengers specifically, and Senegalese society, more generally, for having never encouraged swimming skills. But as this body of work reveals, the forces that shaped the *Joola* disaster and the lack of response by political entities were more complicated than such an explanation alone can suggest.

Those from Casamance argue that it's not simply a challenging marine environment that affects locals' ability to swim in Ziguinchor. It is more than a question of the murky, muck-laden aquatic conditions. Rather, there is a

Figure 3.1 Pirogue Operators in Casamance. Source: Author.

real fear of violence from the resident crocodiles and hippopotamuses that roam the wetlands of the Casamance River and in other riverine areas of the rural south. The hippopotamus is considered the most dangerous mammal in Africa, and in the Casamance community of Colombo alone, far from the coast, twenty-five fishermen were mauled to death by these legendary hippos in one year (Agence France-Presse 2016). Thus while the sea and riverine environments of Senegal provide food security and a space for adventure, the threat of drowning and the documented dangers from marine wildlife present enduring challenges for fisherfolk and potential swimmers (Figure 3.1).

Despite the key role that the sea plays in Senegalese food systems and cultural identity, there is also the reality that there are few *public* spaces in which young people can safely learn to swim. This is especially true in southern Senegal. In Ziguinchor, the capital of the Casamance, there is only one community swimming pool, and when the local pool is not in operation, some use the western style hotel *piscines* as a substitute. One problem is that the entrance fee to swim at Hotels Kadiandoumagne or Aubert can run upward of 3,000 CFA (US$6) per day, making costs prohibitive for large families of limited means (UNDP 1999). In contrast to the beaches of Somone or Saly where swimming is free courtesy of the sea, in the south, the physical geography of Casamance alters the everyday economics of swimming.

Efforts to reduce the high incidence of drownings are underway in the northern coastal city of Saint Louis, and are being driven by civic leaders wishing to change attitudes toward marine space and water safety. These are not top down policies but rather grassroots initiatives. Near the port of Bacre Waly Gueye, Abdou Salam Diop—known locally as Zale—has been teaching local kids to swim for over twenty years. Despite being adjacent to the Senegal River, many people in Saint Louis lack swimming skills, and so Diop initiated a low cost solution. He introduced the jerry-cans (plastic or metal fuel cans) as buoyancy aids in order to teach residents key water skills (Agence France-Presse 2016). Few children, much like *pirogue* fishermen themselves, own safety vests, and so three to four times per week they are taught to jump off docks into the Senegal River and swim with their jerry-cans. In a country where the sea is a central focus of cultural life and fishing one of its primary economic activities, swimming resources and training are surprisingly limited. Thus grassroots efforts have played an important role in solving problems. Locals improvise to ensure their safety at sea, while some just throw caution to the wind and board vessels without life vests. In Somone, I witness a boy using an empty three gallon water (*ndox*, in Wolof) jug as a flotation device in the sea, and by the week's end, I had jerry-rigged two empty bottles to ensure my son's own safety in lieu of a life jacket. Indigenous knowledge of the sea here takes many forms, and today, it is ours to appropriate from those who best understand the Atlantic's risks and rewards.

It is November 2018. It is the middle of the day, the sun high and blazing, and we are huddled on the beach speaking with Muis, who works as a *plunger* (a diver) trapping lobster some 10 kilometers offshore of Somone. We're arranging a trip to free dive for these crustaceans with their long, muscular tails, but mostly to become acquainted with the day-to-day operations of a fishing boat. A few people play in the surf, and then suddenly, the group witnesses Gerard, an apartment security guard, jogging out to the surf in full security gear—black boots, long pants, collared shirt, and baseball cap—before he enters the water to rescue a woman struggling in the waves. Ebo, a restaurant manager, has also witnessed the event, and begins to remove his shoes and roll up his long pants. But it is Gerard who arrives first at the scene, and like a Senegalese superhero, he carries the tourist out of the water until he's able to safely set her down in a chair. Later I learn the story with a mixture of hand gestures and broken Wolof, how Gerard's cell phone is now dead since he didn't have time to remove essentials from pockets before making his rescue. It is a story so fitting and timely given the above account of Senegalese safety at sea. It reveals the complexity and nuance of those locals who are still skilled at reading the ocean. The sea, and Senegal's relationship to it, is a book that I do not yet understand. I will learn later that Gerard and his family are from Ziguinchor, the city that lost the most lives in the *Joola* tragedy.

PIROGUE FISHERMEN AND PLASTIC POLLUTION

We sit on the beach of Somone as the sun sets across the Atlantic, my eight-year-old son using a bean pod to sketch photos into the sand of unfamiliar faces and stray dogs that roam the beach here in this dry, coastal region of the African Sahel. The beach is speckled with trash, despite Senegal's efforts in the 1970s to clean up neighborhoods (particularly in the south) using community competitions accompanied by cash rewards. Here in Somone it is easier to comb the sand for soda bottles, yogurt cups, water sachets, scraps of fishing net, bottle caps, a single red high heel shoe than to locate seashells amid the other ephemera that make its landing on shore. Plastic pollution is particularly high. Some of the men who "do sport" at night on the beach—push-ups, squats, sprints, *futbol*—bring rakes to sweep up the debris. They dig a hole, and then bury the trash adjacent to the crab burrows that pepper the seaside. The Sisyphean task is hard to observe, as each day simply brings more trash, but we join in, rake up plastic scraps and knife handles, my son treating the ephemera like archaeological treasures to add to his arsenal. The beach boys of Somone have begun their own social movement to clean up the sand, and their efforts are contagious.

Two Senegalese *pirogues* are currently anchored offshore, within swimmable distance, our first close up sighting since our arrival here three weeks ago to study the sea and its artisanal fisheries. Artisanal is a word that's been co-opted as a marker for Whole Food cookies or even KitchenAid mixers, but it's true meaning is that of the "human scale," and meant to define those who produce something by traditional methods (Neiland and Bene 2004). The *pirogue*'s anchor lines run from the sea straight up to the seawall, where a crew of sixteen to twenty young fishermen men sit in the sand eating, talking, and engaging in impromptu wrestling, a popular Sereer sport of Senegal (Sene 2016). Behind them is a strip of beach hotels and French vacation homes in various stages of construction, and the young fishers sit with their backs to the seawall, which shows signs of crumbling and decay from the climate induced change that has led to coastal erosion here northward to the UNESCO site of Saint Louis (Breeden 2018). Each night we watch as the high tide erases the last vestiges of the beachfront, waves pounding hotel walls and rock piles, until the tide recedes westward with the moon.

This fishing crew has been at sea for around two weeks, fishing from their wooden *pirogues* in what has been fabled as the richest fishing grounds in the world (Belhabib et al. 2006). Much like coastal South America (northern Chile and southern Peru), in the eastern Atlantic the physical geographic process known as upwelling causes water to rise up from beneath the surface to replace surface water pushed westward by reliable, dominant wind patterns (Capet et al. 2017). Deep subsurface water that rises to the surface of West

Africa is colder, nutrient-rich, and, perhaps most importantly, biologically productive (Capet et al. 2017). In West Africa as in other upwelling regions around the world, the key consequence is a highly productive fishing space that has sustained livelihoods and provided a ready protein source for generations. In a nutshell, you'll find plenty of fish here. Yet this area, rich in biodiversity, has been dramatically altered over time by pollution and increased shipping, and the fishermen here speak freely of this challenge to their lives and livelihoods (Capet et al. 2017).

Anatomy of the Pirogue Fishing Vessel

The *pirogue* is a French word for a small *bateau*, a boat that is handcrafted, and constructed by artisans into a flat-bottomed, banana shaped vessel for use in fishing. *Pirogue*s are found in regions once colonized by France and Spain, and they are typically dugout vessels made from a single log. In Senegal, once part of French West Africa, the FAO estimates that 600,000 people operate these vessels in the coastal waters of the country's exclusive economic zone (Belhabib et al. 2017; Cormier-Salem 1994). Like fishing more generally, *pirogues* form a unique part of Senegal's national identity, and not unlike the colorful fabrics worn by Senegalese women in the streets from Ziguinchor to Dakar, *pirogues* leave an explosion of color upon the landscapes and seascapes, spotted from high above in a plane, or from the beaches below, as *pirogues* ply the waters until arriving onshore (Figure 3.2).

The Senegalese *pirogues* used by fishers are constructed from timber called *khaya*, a genus of seven species of trees in the mahogany family *meliaceae* (Okieimen and Eromosele 1999). Globally *pirogue* timber is known as African mahogany, a species endemic to the tropical belt of Africa but also the island nation of Madagascar. Locals tell me that the mahogany tree or *garapp* (in Wolof) reaches about 35–40 meters in height, sometimes greater, with a trunk diameter that spans no more than one meter (Okieimen and Eromosele 1999). Not surprisingly, *khaya* are planted in some parts of tropical Africa as shade trees in large public spaces and parks because they are fast growing but require fertile, moisture laden soil. Strong, durable, and resistant to termites, the tree weathers well, ideal for the vagaries of the sea, and it is perceived as one of the most important woods in all of Africa. In addition to its physical properties, *khaya* also plays a medicinal role in some traditional African religions, as the bitter bark is utilized to treat colds, and the seed oil sometimes harnessed as a salve for head lice (Doumbouya et al. 2017). It is not unlike the mesquite tree in the American Southwest, multipurpose and sacred, albeit the mesquite is a thorn in the side for ranchers who must compete with the tree's water thirsty nature. While also used for furniture, undeniably the most important use of *khaya* timber in West Africa is to craft the

Figure 3.2 Pirogues at Rest on the Day of Tabaski. Source: Author.

iconic *pirogue*, vessels that feature prominently into the Senegalese resource economy and history.

According to legend, the *khaya* received its name from an African botanist who visited the region centuries ago (Okiemen and Eromosele 1999, 279–280). When he asked for the species' indigenous name, the botanist's guide replied, "khaya," which in the local guide's language meant "I don't know" (Okieimen and Eromosele 1999). The botanist jotted down the name, and the genus has remained in the scientific record ever since as *khaya*. Beyond Senegal's borders, Nigerians have long used African mahogany to create the *dufuna*, known currently as the oldest boat in Africa; research dates its construction back 8,000 years ago. But it is strictly because African mahogany holds a variety of uses in society and maintains durability in contrast to other timber that it has, over time, become an endangered species. The International Union for the Conservation of Nature (IUCN), an affiliate of UNESCO (United Nations Educational and Scientific Cultural Organization) has placed *khaya* on the "red list" of threatened species in the "vulnerable" category, one step below "endangered" on the agency's metric for threatened, endangered, or extinct species (DuBois et al. 2012). It is ironic that in the same way that timber materials used in construction for the pirogues are at risk of extinction, so, too, are the fishing grounds that *pirogues* have fished for centuries.

The making of a *pirogue* is a marvel. The hand tools are simple—hammer, pliers, axe, and handsaw, unlike the industrial mechanization found in other parts of the world. My grandfather, a Norwegian-born U.S. immigrant, spent his life working in the shipyards of Brooklyn's Bethlehem Steel, one of America's oldest companies. He built boats, could swim far and fast, and he lived up to his Norse name of Thor. Ships, boats, canoes, zodiacs, and water, in general, course through our blood. But I find the *pirogue* special because it is fathomable, digestible, existing at a scale that can be fully understood from bow to stern, unlike, say, the *Queen Mary*. Their construction is reminiscent of the fantasy Ghanaian coffins (*abebuu adekai*), crafted in Accra, each with its own unique identity, signifying the deceased person's profession or (e.g., an 8-foot-long camera, for a photographer, or a cacao bean, for a chocolate lover), and meant to accompany the body into the afterlife. They are both built using basic materials but ultimately produce permanent works of art. The coffin is buried terra firma, and the *pirogue*, designed to stay afloat, but they each represent a world made by hand.

I learned that to complete the construction of the *pirogue* tiny cracks are patched with wood-chips or bits of rubber from old flip flops or rubber hoses. An old knot of rope might be used to patch a hole, or anything salvageable. Seams within the boat are often caulked with an emulsion of baobab leaf powder, sawdust, and tar. And once constructed, local artists, or which there are many in the beaches of Somone and Ziguinchor, arrive to paint the *bateaus* shades of green, red, yellow, the national colors of Senegal, though sometimes other hues are used. Naturally, every ship possesses a name, for not unlike other parts of the world it is seen as a sign of bad *jinn* to go without. One friend's *pirogue* (*gall* in Wolof) is assigned the name Fatou, after the captain's mother, fitting since mothers (*yaay)* are designated as the most respected members of Senegalese society.

Pirogues range in size and function and we aim to experience many of them during our stay for a few months in Somone, Ziguinchor, and Saint Louis, each village distinct from one another. Shortly before Tabaski, the national holiday of Senegal, we secure a trip out to sea with Muiss's fellow fishers and two graduate students from France, one generational notch down from us. The boat, smaller than most *pirogues*, is roughly 30 feet long, 3.5 feet wide, contains two crewmen, one at the bow and the other at the helm. The bowman wears shorts, a red windbreaker, and is barefoot, his smile making it clear that there's no other place he'd rather be than here, right now (*leegi leegi*). The helmsman is the same. He is a *bayfall* with shoulder-length dreadlocks, a pair of olive waders, and bare feet.

With the ancient fifteen horse engine—passed down from a different generation—we motor 10 kilometers offshore before the bowman tosses the anchor off the bow and into 15 meters of frothy sea, the anchor fashioned artistically from used rebar. The boat takes on water despite the plugs, and so the *pirogue* men bail, using half of an old antifreeze container to scoop out the murky seawater. The fishers in these *pirogues* fish with lines, not rods, and to do so they wrap a thick, green nylon line around a five by five block of wood, Styrofoam, or anything that floats, really. Four lines are managed at once by the boatmen, the fisher keeping the nylon over his index finger in order to feel the slightest tug from a fish. He begins to pull fish in one after another as if removing socks from a clothesline before a storm. The single onboard knife is absent of a handle, incredibly sharp, and after each use, the fisher sharpens the blade on the edge of the wooden boat, like an onboard kitchen strop. Dorade, lotte, and thiouf—local staples in Senegalese restaurants—are caught quickly, in succession, ranging in size from just 5 inches to twelve. *There are plenty of fish in the sea*, so goes the American idiom, but they are soberingly small six miles off Senegal's coast. To be fair these artisanal fisherfolk are not the culprits in this region's ongoing marine degradation. They take the "small-small" which does not benefit the food chain but they are forced to take what remains or starve (Du Bois and Zografos 2012; Evans 2005).

THE SEA AND THE SUPERNATURAL

The *pirogues* can be much larger, but regardless of the vessel's size, most fishermen engage in rituals to encourage a good harvest, protect themselves at sea, or keep *jinn* from stealing schools of Atlantic fish. Miseur Omar Sene, a former military man and the father of Djiembe, my son's Wolof tutor, has lived in the Ngaparou area for over thirty years. Two of his young sons are fishermen, and on the holy day of Tabaski we sit at the beach under his family's palapa hut in plastic chairs, discussing the sea and the power of mysticism (Figure 3.3). He has rolled out his woven mat in anticipation of the afternoon call to prayer. I ask him about *jinn*, since we've witnessed different types of rituals preceding a *pirogue*'s departure into the surf. I inquire as to whether these rituals are an adaptation to changing economic and environmental realities of the sea. Is this related to climate change, or piracy? He counters that such rituals have been present since Casamance was part of the Mali Empire. Sene describes how, for instance, the *Joola* shipwreck was both a political event and a mystical one, a notion that few mentioned in any of my interviews. Most respondents commonly cited the political crime of the *Joola* tragedy, but quietly, Sene says to me: "There was an animal, a seven headed snake known as *ningkiauauka*, who consumed the overloaded *Joola* as the night's storm approached. Many believe the sea monster sank the ship."[1] Omar Sene is an educated man, but in his remembrance of the shipwreck he clearly believes in the power of *jinn* alongside his adherence to the religious tenets of Islam. I ask if the *Joola* was blessed by marabouts in advance of any of its journeys, especially on September 26, as the *pirogues* often are, but the question is left unanswered and we fall silent in the hot afternoon sun.

I tell Miseur Sene about the first time we witnessed two large *pirogues* launch off Somone at high tide, the sun setting, Senegalese families out on their evening walks, men doing push-ups on the beach. Perhaps twenty fishers climbed aboard, maybe 15 percent of them donning what appeared to be bright new orange life jackets. I was told by Mustapha, a forty-four-year-old man who sells tourist bric a brac on the beaches, that the lifejackets are now a new element of *pirogue* culture. They were mandated by the government after the *Joola* wrecked, but most fishers rely less on the jackets and more on the power of marabouts and *gris-gris* to ward off bad luck (Villalon 2004). Laws, as the saying goes, do not make people behave. *Gris-gris* really comes down to the simplicity of Quranic verses wrapped in cloth, then tied to a boat or placed within one to ward off *jinn*, the bad spirits who may steal fish or make mischief. The power of tradition and mysticism is everywhere but most evident on some of the larger *pirogues* where fisherfolks may be gone for two weeks at a time (Figure 3.4).

With the smaller *pirogues*, we watch the fishermen drag the boats from the sand into the rising tide, rolling the hulls across mahogany logs to facilitate ease of movement into the surf. But with the large *pirogue*, the boat stays anchored offshore. Men then board the vessel and pull hard on the anchor line fist over fist until they reach the anchor, while simultaneously gunning

Figure 3.3 Friends in Ngaparou on Tabaski. Source: Author.

the engine. As they do this they chant "strength, strength, strength," multiple times, jump up and down, while keeping perfect balance while standing onboard the vessels. They are the yogis of the sea, and possess a core so strong both in spirit and physical form that borders on the supernatural. Some perform rituals before they depart, and Sene explains that the mysticism of *jinn* infiltrates all aspects of Senegalese life not just those who draw their livelihoods from the sea.

The belief in *jinn* is common regardless of the age-set within Senegalese society. "Young fishermen," Sene says, "perform sacrifices to wake up the boat, since it is considered asleep when it is not at sea. The sacrifices are sacred, indigenous knowledge. Verses from the Koran are written in the sand with the index finger, the sand is picked up, and then sprinkled upon the *pirogue*'s bow. A marabout may bring *gris-gris* to bless the boat, or in cases where fishing luck has been poor, a goat may be sacrificed." Sene draws a star in the sand, and at each point, pulls a few grains to show a sample of how the ritual unfolds. I ask if this information is meant to be shared, and he smiles, saying that "it is generally sacred, but that these rituals are really not complicated." Protecting a boat against *jinn* may be as simple as pouring a glass of milk on the bow, or a marabout saying a small prayer. It is interesting that non-Murids[2] who operate within the Sufi world have also adopted traditions to ward off spirits and encourage good luck while fishing. Seated in his white traditional robes, Sene tells the story of a Spanish captain of a Senegalese *pirogue* who adopted the custom of placing a full bottle of water in a purse seine before each catch.[3] "Every time he dropped the bottle, he received a reward from the sea—a full net of mackerel." This particular captain was a practicing Catholic and not a Sufi, more evidence of the mixed religious traditions of Senegal.

I mention the American freemasons and their own signs and shared internal codes, as my father was a lifelong mason, a world that was kept secret. We witnessed firsthand the funerary freemason rights at my father's funeral in New York. Sene and others explain to me that freemasonry became common first in Nigeria, and then in Senegal, under then president Abdoulaye Wade. I ask Sene if he sees parallels between the marabout's signs and the masonic rituals, or if any of freemason's principles have been applied to marine space, such as whether masonic rites are used on *pirogues*. He does not answer the question but mentions how freemasons have tremendous power and financial means here in Senegal. "It is different with *jinn*," he explains. "African beliefs about the sea and protection from its *jinn* are compatible with Islam, and sometimes intertwined in their everyday execution. You can be from the Lebu ethnic group like me and my family here in Ngaparou but you can also be a priest, or you can follow Islam. You can pour milk on your son or porridge on his boat to calm either one, but you can still participate in the evening call to prayer."[4] Though this may sound much like the fusion of traditions in the Americas—where Old World religious customs were combined with indigenous American beliefs—many West African scholars do not like to employ the word *syncretic*. To them, the word implies that one culture has subsumed the other. Rather, this fusion of everyday practices and beliefs is simply a way of life here in Senegal, what the Pew Foundation called in a 2016 study "peaceful coexistence" along a religious fault zone in Africa (Pew Report 2010). The conversation with Omar comes to an organic end, because soon a plate of mutton is brought out to our table under the palapa hut for us to share. It is clear that on this day of Tabaski there is time to talk, but also time reserved to eat and pray.

FISHERMEN AS FIRST RESPONDERS AND HEROES

In November my friend Mbaye Ngom in Casamance brings me past the familiar neighborhood of Tildene Kadior to introduce me to Major Diatta, a former military man and well-respected citizen of Ziguinchor who lost his nineteen-year-old son on the *Joola*. Together Mbaye and I enter the compound and are greeted by Major's family and friends, followed by the exchange of traditional Senegalese greetings. Mbaye wears a cowboy hat, as many do here in the rural south, quite different in cultural composition than the urban north or coastal regions. His cowboy hat is weathered, much like mine, but it is also American made—a Stetson—whereas ironically my hat is a knock-off crafted in China and bought in Alaska. Major Diatta, the patriarch of this compound, exudes a beautiful tranquility, and proceeds to show us the image of his son Abdoulaye, a photo which adorns the wall adjacent to his own military mementos and family images. Here there are certificates of recognition and a sepia portrait of his own father in full Wolof traditional dress. Major Diatta clutches handwritten notes and asks me to sit, evidence that he has prepared for this meeting well in advance. He personifies respect in his style and dress, his overall comportment with the world. His notes are in French cursive, and without error, they underscore his attention to detail, the legacy of a life as a military man.

I am instantly at peace, seated here with Major and his best friend, a silent companion, holding his prayer beads. The Major is ready to talk about his son and that day in September, now some sixteen years after the ferry wreck. In the next few hours he will explain how his boy Abdoulaye was an expert at taekwondo, how he wanted to be a doctor, but before that he leans in calmly and whispers in French: "The *pirogues*. They arrived first at the scene." "The *pirogues*?" I ask. "*Waaw*, he responds, *le pirogues*."[5] The fishermen pulled bodies from the water and the pirogues were the first responders. Major is

Figure 3.4 Fish market in Senegal. Source: Author.

tall and lean, in his sixties, and in traditional clothing exudes the energy of a spiritual leader rather than the archetype of a military member in southern Africa. "No one else came to help the *Joola*," he says softly. These words, spoken from a man who worked for the government for the majority of his life, is the first of many unveiled criticisms I will hear of what Senegalese fearlessly call "an avoidable tragedy."

After the shipwreck, the Major took it upon himself to become an activist in Senegal. He spoke publicly about the *Joola* event, and his gravitas as well as his role as a respected military man led people to listen. Two of the cemeteries that were built to honor the dead, most of them populated with victims whose names were never identified but numbered, are maintained exclusively by Major and his family. As we speak in his compound, women of all ages drift in and out of the room to unknown destinations, offering up porridge or tea to Mbaye, Major, and me. We discuss more, and I learn that the Major has donated his financial and temporal resources to paint stone cemetery walls, maintain grave markers, and help those families who lost daughters or sons to the shipwreck. In our conversation we keep circling back to the *pirogues*, and his name for the artisanal fishermen: *the first responders*. Chapter 4 will outline in detail the role that fishermen played as first responders during rescue efforts of the capsized passengers. Yet for now, any discussion of Senegalese sea space requires at least some mention of these human agents of change and resilience.

Fishing Piracy and Senegal's Sea Spaces

One prominent issue that surfaced in interviews was the way in which fishing piracy has undermined the everyday livelihoods of the *pirogue* operators outlined above. The presence of these illegal vessels also represents a physical incursion into customary sea spaces of the region. Thus, the heroes of the *Joola* wreck—*the first responders*—are now being threatened by economic forces operating unlawfully within their maritime territories. In fact, many interviewees described piracy as the greatest threat facing Senegal since the *Joola* disaster and international research supports this local observation. Illegal, unreported and unregulated fishing (IUU) has become a crisis of environmental and cultural sustainability in West Africa. Current rates of marine extraction in Senegal, Ghana, Sierra Leone, Liberia and Mauritania (West Africa) are driving key fish species toward extinction, and the abuse of marine resources is a product of larger piracy forces, not a function of the traditional *pirogues* that ply the coastal inshore waters (Belhabib et al. 2006). The National Oceanic and Atmospheric Administration (NOAA) also recognizes that IUU fishing is a global problem threatening not only ocean ecosystems but also local livelihoods (FAO 2008). Many West Africans, as noted above, rely upon fish as the sole source of protein and a marker of cultural identity.

Why Senegal? Due to favorable natural conditions the country boasts some of the richest fishing grounds in Africa, accounting for 75 percent of the state's animal protein consumption (Summers 2019). As detailed earlier a total of 600,000 artisanal fishermen operate lines, traps, and nets, representing 70 percent of the country's total catch volume (FAO 2015). Yet IUU vessels from China and Russia who have exhausted much of their own piscatorial base now cull Senegal's exclusive economic zone (EEZ) for protein. In the process these vessels violate marine conservation measures that were set in place through international agreements (Belhabib et al. 2016). In one year a Chinese mega-trawler may capture 20,000 tons of fish, an amount equal to the annual take of 1,700 Senegalese pirogues (Faye 2011). Ample research has investigated the large-scale impacts that IUU incursions have on fish stocks, but there has been scant research into how communities *resist* or *react to* changing circumstances.

Political Ecology of Sea Space in the Developing World

In this vein Nietschmann (1994, 1995) was actually the first to unravel the way in which political ecological forces such as piracy shaped the lives and livelihoods of fishing communities in the developing world. His geographic fieldwork along Nicaragua's Miskito coast explored how foreign demands for *Caribbean red gold* (the local name for snapper, lobster, and shrimp) led to marine degradation. IUU fishing from nearby Panama, Jamaica, and Honduras had nearly decimated Nicaragua's fish stocks in the 1970s–1990s, and in response, the Miskitos proposed creating a Marine Coast Protected Area to safeguard their traditional marine spaces. Miskito Indians also took up arms—specifically, AK-47s left behind after the Sandinista-Contra War—in order to patrol their coast and demarcate the extent of their indigenous resource base (Nietschmann 1995). Many lessons on community resilience and resistance can be drawn from the Nicaragua case study for the Senegalese maritime context.

Piracy and Migration

From a political ecological perspective resource piracy has not only altered the economic livelihoods of *pirogue* fishermen in Senegal but driven migration patterns within these communities. A recent study indicated that IUU fishing costs West African (Senegal, The Gambia, Guinea-Bissau) economies some US$2.3 billion a year in lost income (FAO 2008). Due to these incursions, artisanal fish catches in the seaside community of Joal alone dropped from 150,000 tons to 90,000 tons between 2015 and 2016 (FAO 2015). As fishing decreases as a primary source of protein, coastal communities have invoked a reverse migration pattern, heading toward interior forests in order to pursue farming activities (Weissman 1990; Wanyoni et al. 2016).Yet reduced rainfall due to climate change has led to periodic droughts, challenging those seeking work on interior farms and fields. As an example, mango production, a key subsistence economy in Casamance, has seen a 70–80 percent decrease in production due to warmer temperatures that began in the 2010s (Dinar et al. 2008). There have thus been domino effects that IUU piracy has played on migration, as communities from "fished out" regions move their families inland to become farmers.

The *pirogue* itself has also become a vehicle of migration, as more and more Senegalese, faced with a shortage of jobs, brave the high seas to emigrate to other countries for work. The *pirogue* is viewed as a vehicle for social mobility in a country where the jobless rate is 15.7 percent and the poverty rate, 38 percent (World Bank 2019). In the 2013 film *La Pirogue*, Senegalese director Moussa Toure tells the story of Baye Laye, a boat captain in a city outside of Dakar, who transported thirty men to Spain's Canary Islands in search of better lives. Toure described how the film was inspired from his own experience, in which nearly every family he knows had a member who had used a *pirogue* to emigrate to Europe. The shortage of jobs on *terra firma*, in addition to the impacts at sea by way of resource piracy and the effects of climate change, are altering the everyday livelihood patterns of Senegalese, not to mention the sociological structure of families, who are forced to migrate to support families.

FOLLOWING THE FISH TRAIL OUT OF SENEGAL'S SEA SPACES

Who is extracting fish from Senegal's EEZ, and upon whose dinner plates do these fish—dorade, capitaine, mackerel, sardines—ultimately end up? According to the Food and Agriculture Organization, approximately 40 percent of Senegalese fish will be sold (or served) abroad, from China to Europe to Russia to the United States (FAO 2008; Jacobs 2017). Evidence near my research site of Somone illustrates the Chinese, Korean, and Spanish presence in Senegal, where fish-processing plants have been established within villages. These industrial plants have appeared all along the Senegalese coast, as well as The Gambia and Mauritania, and are not unlike the resource extractive outposts that enriched Europe's coffers at the time of colonialism. Gaossou Gueye, president of the Confederation of Traditional Fishermen in West Africa, explained: "The demand for fodder fish to feed salmon and carp has generated an extremely voracious industry that has monopolized fishing resources, because it takes five tonnes of sardinella to produce a single tonne of fish" (Summers 2019). These fishmeal plants are lucrative endeavors, and can pay as much as 4,000 West African CFA for 50 kilos of fish. Their product is then exported to China, Korea and the European Union (Senegal Data Portal 2020).

Along the coast of Senegal, fishmeal exports rose from 990 tons in 2006 to 6,288 tons (the equivalent of 31,440 tons of fresh fish) in 2015. Just a few years ago, one tonne of fishmeal was worth between $1,200 and $1,500, and now that same quantity is valued at $1,800–$2,100 (Summers 2019). Not surprisingly, local fishing communities have not profited from the windfall, and women who work not at sea as *pirogue* operators but as fishmongers have been pushed out of the market. According to Marieme of Somone: "We just can't compete with factory prices and demand." Fishermen recognize that the federal government has done less to support artisanal pirogues and more to benefit global actors. Local fisherman and conservationist Abdou Karim Sall stated: "Instead of fulfilling the needs of a population whose employment and food security depends on fish, Senegal has opted to invest in partially government-owned entities and industrial fish processing plants that are financed by the Koreans, Chinese or Spanish, who in turn benefit from looser regulations because their boats are registered under the Senegalese flag" (Summers 2019; World Food Programme 2011).

While in Dakar, I witnessed firsthand Chinese vessels flying Senegalese flags, anchored at an independent port beyond the capital's maritime hub. This port had its own security force and even operates outside the control of Senegalese authorities. Such activities—both the fishing extraction as well as these neocolonial entities—threaten the livelihood and power structure of the entire region. *Pirogue* fishermen depend on sardinella for jobs. And because sardinella are at the bottom of the food chain that sustains other species, their extraction by global actors is having a domino effect upon other keystone species and the lives of fishermen and women. As stated earlier in this chapter, the life of a *pirogue* fishermen features prominently into Senegalese identity.

In addition to fish-processing plants owned by non-nationals, I am told that reefers have had one of the most significant impacts upon artisanal fishers in Senegalese and other West African communities. A reefer is a large-scale commercial vessel that receives and freezes fish at sea—and uses containers for shipping. Recent research provides evidence that these reefers undermine multilateral governance rules aimed at curtailing IUU fishing and promoting sustainable, legal practices. Senegal has taken the lead alongside the nation of Cote d'Ivoire to ban what is called "transshipment" in their waters and other West African governments would benefit from the same (Wanyoni et al. 2016). The FAO recommends that ships caught engaging in IUU fishing should be blacklisted and prohibited from entering waters in the region. But these are merely recommendations without political or economic teeth. Much like the *Joola* shipwreck, local actors are seeking justice in the face of a tepid government response.

RAFET NJORT AND COMMUNITY RESILIENCE IN THE FACE OF PIRACY

Few scholars beyond Nietschmann (1995) have emphasized the human agency of fisherfolk in terms of their adaptation, resistance, and resilience. Political ecology laid the foundations for *community resilience*, defined as "a measure of the sustained ability of a community to utilize available resources to respond to, withstand, and recover from adverse situations." Senegalese *pirogue* fishers have done their best to adapt to changing circumstances along the coast, though their tactics differ from those of Nicaragua's lobster fishermen who patrolled their coast with AK-47s left behind by the Soviet regime (Nietschmann 1994). In Senegal, fishers are instead forming local cooperatives, working to protect marine spaces in the form of maritime reserves and quite simply, engaging in the Sufi rituals and beliefs to guide their everyday lives. Likewise, in the fishing and farming sectors of Ziguinchor, locals have enacted food for work projects—restoring mangroves and building salt blocking dams—which illustrates local responses to climate change, too.

Clearly, multiple global forces coupled with an anemic national response were evidenced during my fieldwork in Senegal. What frustrates most *pirogue* fishermen I talked with in Somone, Saint Louis, and Ziguinchor was the lack of support from the government. Indeed, they sometimes have skirmishes at sea with foreign trawlers, such as the case when purse seine nets intercept smaller, artisanal nylon nets (which are technically illegal but still used). Yet fishermen spoke of the powerlessness they feel not just against the pirate vessels hoovering away sardines, but in the face of the government and its failure to support artisanal fishing: the heart and soul of Senegal's identity. *Pirogue* fishermen possess an agency that enables them to navigate the everyday vagaries of the sea. Their job is harder than just about any I've seen to date, save that of the slaughterhouse workers who toil in the JBS meat packing facility in Colorado, many of them African immigrants from Somalia (Barton 2017). The point is that *pirogue* operators respond to these incursions in the way they know best, but a large proportion of their energies are funneled into the everyday, rigorous process of working at sea. The fight against the fate of the fisheries is often too great to bear, and regional cooperation is needed in the wake of unchecked IUU activities (Bortei-Doku Aryeetey 2002). Most importantly, the struggles facing *pirogue* operators echo the lack of justice facing survivors and families of the *Joola* disaster.

I conclude here that Senegalese optimism may be seen as a form of community resistance and resilience. *Rafet njort* is a Sufi principle taken to mean "beautiful optimism" or "good intentions" and it infuses the way in which Senegalese navigate the possibilities and perils of everyday life (Ngom 2016). In my four visits to Senegal spanning three years, it was rare to encounter an individual who maintained a negative outlook on life. Ups and downs were common, but the general sensibility, guided by Islam but also dominant cultural norms, is that "tomorrow will always be a better day." Likewise, young fisherfolk view the fate of fishing through the lens of *rafet njort*, but with factors that are complicated by age-set and livelihood experiences. In the United States, the idiom goes, "Nothing good worth having comes easy." In Wolof, the proverb suggests, "Ku bëgg akara dangayñeme kaani," or whoever wants accra[6] must brave the peppers. As one elder explained, you have to get past the spice of kaani to reach the sweetness of the accra.

In a group interview with fishing communities in Ziguinchor, men in their fifties described a time period in which pirogues would sail for only a few hours before the crew would spot a school of fish. The men would cast once or twice with their purse seiners and haul a full boat of fish—1,000 kilos or more, allowing the team to return to shore by lunch. I was told this same story by multiple seniors from Ziguinchor to Somone to Saint Louis, and while one never knows the difference between fact and fiction when it comes to fish stories, nostalgia was the common theme. In contrast, fishermen of the younger generation argued instead that the sea was not empty of fish but that it "simply came down to optimism and a little good luck."

In the context of resilience it's worth noting that despite sobering statistics about African poverty in general, Senegal in particular remains an anomaly in terms of subsistence affluence. It lies at the western edge of the Sahel, south of the Sahara desert and in the middle of a geographic belt that the Pew Foundation calls a religious fault zone (Pew Report 2010). Much like the news media in its portrayal of the Sahel region, geography textbooks also perpetuate the image of a place ridden with perpetual mass hunger and desertification-induced cattle deaths. Indeed, Senegal is a poor country, but in every interview held during my years in the country, people repeated the same mantra: "We have many problems, but there's always enough to eat." I would hear how: "We can pull mangos or soursop from trees" (Figure 3.5) and I was told repeatedly that "when the rains arrive in Casamance, the streets are flooded with fish." This notion of subsistence affluence is especially true in coastal fishing communities, where, despite the hoovering away of resources by the Global North, and the sea's emptying out, people see themselves as having plenty of fish.

Figure 3.5 Mango market. Source: Author.

Every day fishers here dive for lobster in rocks just offshore Somone. They head 10 kilometers into the sea from Saint Louis to catch capitaine, a fish that tastes as light as air. They scour the inshore of Cap Skirring for sea urchins, or send pirogues out for squid, offering up small squishy bits to one another during beach workouts. They speak of *ceebu jenn*—the national dish—as if it is a religion, as if fish and rice have the power to heal in the manner of a marabout. The Senegalese coast has plenty of fish, though the truth is, they are getting a lot smaller. Regardless of these political ecological changes, we must remember how the space of the sea remains a marker of Senegalese identity and a source of cultural pride. There have been countless losses at sea on *pirogues*, as there have been drownings due to the issue of safety discussed earlier in this chapter. But the *Joola* shipwreck, the most significant disaster to ever occur in West African marine space, challenged a nation's identity, unified *pirogue* fishermen as first responders, and launched a conversation about how best to manage this common property resource in the face of new global threats: piracy and climate change. Senegalese fishermen continue to take on the mantle of first responder in multifaceted ways guided in part by their own religious tenets, optimism, and resilience.

NOTES

1. Interview with Miseur Sene, Ngaparou, Senegal, November 2018.

2. In Sufism, a *murīd* (Arabic: ديرُم, literally "one who seeks") is a novice committed to spiritual enlightenment by *sulūk* (traversing a path) under a spiritual guide (Ngom 2016).

3. A purse seine is a type of fishing net that hangs vertically in the water with its bottom edge held down by weights and its top buoyed by floats. They can be deployed from shore as a beach seine or from a vessel (Marine Stewardship Council 2020).

4. Interview with Miseur Sene, Ngaparou, November 2018.

5. Interview with Major Alpha Diatta, Ziguinchor, November 2018.

6. A fritter made with black-eyed peas or a similar pulse, or, especially in Trinidad, mashed fish.

Chapter 4

Shipwreck

An Accumulation of Errors

Whatever is painful does not last, and whoever perseveres smiles.[1]

Wolof proverb

While hundreds of wrecks have been documented in coastal maritime spaces in Africa dating back to 1747—including convict ships, foreign patrol boats, merchant vessels, fishing trawlers, and slave ships—no vessel's fate is more remarkable than that of the *Joola* passenger ship (Figure 4.1). Pat Wiley, who has published the only *American* journalistic account of the *Joola*, famously described the ship as "Africa's Titanic," a moniker that despite its accuracy has caused some Senegalese to bristle (Becker 2012; Wiley 2012). In the words of a colleague from Casamance: "It is regretful that outsiders beyond France know little of the *Joola* wreck. Sadder still is that to get any attention the *Joola* must be compared with, and judged by, the event in relation to the America-bound Titanic."[2] The *Joola* killed an estimated 1,863 people and left only sixty-four survivors, and while the wreck created more casualties than the *Titanic* there has been little coverage of the story nor an investigation into the people and places of the *Joola* (Barton 2017). While there are numerous French publications on the subject, Wiley's (2012) account is one of the only existing and proper treatments of the events leading up to the tragedy.

Now that we have sufficiently laid the foundations of Senegal's colonial cartographies (chapter 1), natural resources (chapter 2), and maritime geographies (chapter 3), we can turn to the *Joola* narrative itself and the aftermath. This chapter contains information on the *Joola* ferry, its route, the people of the ferry, and the multiple, intersecting, and alleged causes of the wreck. It highlights those who were most affected and why, and discusses how locals including the *rescapes* (survivors) have responded since the disaster. For those readers interested in the gore of the disaster, they should turn elsewhere to other news stories that outline these narratives. This project is about people and places, the forces that shaped the tragedy, and the ongoing narrative of resilience and the local desire for reckoning. Using interview data I explore how Senegalese citizens managed to move forward in the wake of Africa's greatest maritime tragedy through a series of culturally grounded strategies. This chapter attempts to fill this existing gap and explore the spatial dimensions that led to the crisis and its aftermath.

In developing this project I spoke frequently with students in my upper-division Africa course about the *Joola*, and it is no exaggeration to say that none had heard of the event, whether in other classes or in mainstream literature. I also attended an Africa Colloquium at the National Autonomous University of Mexico in October 2019, and two professors from South Africa noted that the *Joola* crisis did not make their country's news cycles. The same was true for my students, who were disappointed to have learned this piece of history for the first time. One undergraduate in my Africa course reported feeling cheated by history, and another wrote how "this is just another example of how Africa has been ignored and undervalued in the news, but also in terms of our education."[3] In class we often tied this problem to the larger notion of Africa as a perceived site of poverty-disease-warfare, one which excluded not just events but the underlying narratives related to such stories. It was enough

Figure 4.1 ***Le Joola in* 1995.** Source: Carol Keicker.

to have history ignore the *Joola* tragedy as just one more horrific event in Africa's perceived trajectory of death and loss, but in the process, the lives and narratives of those involved were also erased (Barton 2018). Students talked about feeling duty-bound to amend this history of omission, but were not sure how to proceed without reinforcing the actual stereotype of Africa as a site of tragedy. That is why this book is less about the event, and more about the people and spatial interactions that gave rise to the day.

Some might argue that there has been such an immense quantity of shipwrecks in Africa that many readers have chosen to ignore them in the context of larger world events. Yet seen in another way these disasters and our responses to them reflect our larger attitude toward Africa in general—one, as George Kimble noted, of general ignorance of the continent (Kimble 1951, 157). In the words of Howard French, "for many Americans, the 'real Africa' is a blurred concatenation of game parks, starving infants and genocidal warfare—or it's a Disney-fied cradle of civilization" (French 1994, 2). Yet stereotyping of Africans has *not* always existed on this scale, evidenced by mid-sixteenth-century European images of Mansa Musa in which African monarchs were perceived as powerful and equal to European kings (although Musa was in fact more powerful).[4]

In terms of the sheer number of deaths from the *Joola* wreck, the story surpasses the *Titanic* and given those metrics is worthy of deeper understanding. Likewise, Senegalese citizens know the story of the *Titanic*, yet Americans and most Europeans have no knowledge of Africa's *Titanic*, nor of the people most impacted. As *Joola* survivor Ousmane Keita stated, "Most of us had grown up learning about the *Titanic*, and the horror of it crossed our mind whenever we boarded the ferry."[5] Thus the focus of this book is on the communities of the *Joola*, and the goal of this chapter is to tell the story of the shipwreck as a way of appropriately honoring the people who were most affected. Out of respect for the victims and their families, this is not a disaster narrative, but rather one that outlines the geographic facts of the crisis.

AFRICA'S FERRY WRECKS

In the case of Africa there have been a multitude of ship and ferry disasters including those occurring off the coast of West Africa. Some of these make for fascinating shipwreck stories, such as the *Diemermeer* in 1747, an East Indiaman that wrecked off the Banana Islands in Sierra Leone. The wreck was left undiscovered until 2014. There is also the *William Lawrence*, a full rigged ship built in Nova Scotia, which sank off the coast of Dakar in the late eighteenth century. And, on July 2, 1816, the *Medusa*—a French passenger ship—wrecked near the Bank of Arguin off the west coast of Mauritania between Nouakchot and Kouadhibout. The traumatic experience of the passengers and crew was immortalized in Gericault's painting entitled *The Raft of the Medusa*. This artwork depicts one moment in the wreck of the frigate *Medusa*, shortly after it ran aground in what is now Mauritania. Following the ship's sinking, at least 147 passengers were set adrift on a hastily constructed raft and all but fifteen people died at sea two weeks before their rescue. For those who survived starvation, dehydration, and cannibalism, they remember the scandal in France that ensued, given the incompetence of the *Medusa's* captain. The *Medusa* story, because it contained so many French passengers, was shared with the mainstream media at that time. Yet many other shipwreck stories remain untold, and as one Senegalese colleague suggested, this may be "because there were largely brown or black bodies in the water, not Europeans or Americans."[6]

Like many African nations, Tanzania in east Africa has seen multiple ferry wrecks in its history, several of which have involved overcrowding like the case of the *Joola* (Burke 2018). In 1996, the MV *Bukoba* capsized in Lake Victoria killing 894, making it one of the worst disasters of the twentieth century. In 2011, the MV *Spice Islander* 1 sank off the coast of Zanzibar leaving around 200 victims.[7] In 2012, at least 145 people died on the MV *Skagit* when a packed ferry sank while transporting people to the island of Zanzibar in the Indian Ocean (Burke 2018). Most recently, in 2018, the MV *Nyerere* capsized close to Ukara Island on Lake Victoria; when it overturned on its way to Bugorora it left 207 people dead. Like the *Joola* these cases involved severe overcrowding, poor ferry conditions, and a tepid government response. In the wake of the MV *Nyerere* tragedy, Tanzanian president Magufuli arrested the ferry managers and declared four days of national mourning (Burke 2018). Yet the opposition party criticized Magufuli's response, stating that concerns about the ferry's condition had been repeatedly raised. Similar to the case of the *Joola*, the Tanzanian government had turned a deaf ear on these community-based warnings related to safety. The disasters in Tanzania possess enormous similarities to the case of the *Joola*, and much like it, few outside the region are familiar with the impacts it had on African society and space.

Scholars like archaeologist Ibrahima Thiaw have worked diligently to give more attention to the human toll of shipwrecks in West Africa, but with a primary focus on the transatlantic slave trade (Thiaw 2003, 2011). The western edge of mainland Africa contains some of the most important wrecks of that time period, in which African men, women, and children bound for America sank on "coffin ships" as they were ferried to the Americas. Yet despite the enormous interest in this historical period, few wrecks have ever been excavated because they were considered too difficult to find, and the work seen as cost prohibitive (Wiley 2012).[8] Thiaw has argued that slave wrecks are also considered a local source of shame, and in some cases not deemed worthy of detailed investigation. As a result of this perception, the only known excavation of a slave ship that sank off Africa while carrying slaves is that of *São José*, which lies off the coast of South Africa. It is the *São José*'s artifacts alone that have been featured at the new Smithsonian National Museum of African History and Culture, and scholars like Thiaw hope that new shipwreck discoveries related to human trafficking will someday be recognized for their rightful place in African history and remembrance (Sieff 2020). It is my hope that someday the *Joola* story will also be presented in such museums in order to expose African narratives to American and European audiences. In the spirit of that objective this chapter now turns to the *Joola* narrative, beginning with the ferry and its construction.

THE MV JOOLA AND ITS SPECIFICATIONS

Built in Germany for the Republic of Senegal, the MV *Joola* was employed on the Dakar to Ziguinchor coastal route and operated by the Armed Forces beginning in 1990. The *Joola* was a 2,087-ton roll-on roll-off ferry (length: 79.5m; beam: 12m; draft: 3.1m) with a deadweight capacity of 500 tons (Alchetron 2018; Thomas and Tricot 2008, 56). In maritime parlance deadweight is defined as the "maximum weight a ship is designed to safely carry when fully loaded with crew, passengers, cargo, fuel, water and stores" (Wiley 2012, 17). The distribution of that deadweight was critical to the stability of the ship, and most importantly, to the safety of the *Joola* passengers. In cases where fuel, water, and cargo were properly stored, these factors contributed to positive stability. In instances where passengers and crew were located higher up above the water line, negative stability ensued. There is ample evidence that negative stability was in place on September 26, 2002, when more than 1,000 passengers beyond capacity loaded the ship that morning, many of them flocking to the top deck to escape the African monsoon heat, and in the process, further enhancing the ship's listing to port (Ewens 2002).

In addition to the deadweight limitations, the *Joola* ro-ro ferry had an official carrying capacity of forty-four crew members, 536 passengers, and thirty-five cars (Technical Commission of Inquiry 2002). For those unfamiliar with maritime terminology, roll-on/roll-off ships (also known as ro-ros) are designed to carry wheeled cargo including cars, trucks, semi-trucks, trailers, or even railroad cars. Vehicles must be driven on and off the ship using built-in or shore-based ramps, which can be located in the stern, bow, or even the side of the ferry.[9] Since its inception the *Joola* ferry loaded vehicles and passengers from the stern with the exception of those who boarded from Carabane Island, which at the time had no formal point of disembarkation other than *pirogues* that would rendezvous with the ship on water and deposit passengers. The MV *Sewol*—which went down en route to Incheon, South Korea—was also a passenger/ro-ro ferry (BBC 2017). That tragedy led to the death of 476 people, mostly secondary school students from Danwon High School in April 2014 (Kim 2017). In this case, the Korea Maritime Safety Tribunal identified three main causes of the accident: modifications to the ferry, overloading of cargo, and drainage of the required ballast water, which led to the ship's sinking (IFRC 2002; Sang-Hun 2019). Much like the *Joola* narrative, the *Sewol* tragedy had drastic implications for young people who were disproportionately represented onboard (Lavery 2019).[10]

The *Joola* was built to Senegalese specifications with the intent of transporting passengers and vehicles in *both* rivers and Atlantic coastal areas (Godard and Teurnier 1992). This dual mandate was part of the problem, in that the ship had to maneuver two very different aquatic environments. The ferry had to be capable of navigating the shallow waters of the Casamance River as well as the deep-sea passage of 150 miles between the ports of Dakar and Ziguinchor. As a consequence of the need for river travel the ship's shallow-draft, nonexistent keel, and flat-bottom hull had been designed for a depth of only 12 feet for the submerged portion of the ship (Wiley 2012, 17).[11] Her seagoing stability was also dependent on internal ballast of fuel and water, for which the flat-bottom hull and shallow-draft design did not facilitate (Becker 2012). With two water tanks both located on the ship in its central sector, up to 125 tons of ballast could be added assuming they were filled, but the captain had no mechanism to trim the ship or offset any list (Auvray 2012).

As for the fuel tanks, they were able to store 160 tons of fuel; these tanks were asymmetric on the port and starboard side, so a transfer pump was employed to trim the ballast should there be a list while at sea. The problem was that the pump was too small to carry out the transfer in a quick and efficient manner should problems arise. Given how quickly the *Joola* capsized, such a maneuver was impracticable had it even been possible. Also worth noting is that ro-ro ferries top off fuel and water prior to departure, but this did not happen in Ziguinchor before the *Joola* disembarked. The water levels were not adjusted due to low pressure issues with the pump (Thomas and Tricot 2008, 45–51). And in terms of fuel, the city of Ziguinchor, unlike Dakar, did not have sufficient resources to refuel a ship as large as the *Joola* back in 2002 (Becker 2012). Instead the protocol was to refuel once the ship reached Dakar, thus readying the ferry for its roundtrip journey. The port of Ziguinchor was in some ways more a pit stop than a destination, although most communities from Casamance considered the ferry their own. These issues of ballast led to the boat's troubling, visible list even as it left the port of Ziguinchor, something many people commented on as they saw their friends and family board the ship on September 26 (Jullien 2012).

There were other widespread concerns with the ship including ongoing mechanical problems and consistent criticisms of her geographical limitations. After the wreck the shipbuilder Neue Germersheimer Schiffswerft claimed that the boat was designed solely as a coastal vessel, meaning she should sail no more than six hours (roughly fifty nautical miles) from shore. Other experts argued that the *Joola* was only intended for river waterways such as the Rhine or the Casamance River, though the Senegalese public was never made aware of these spatial restrictions. Also, with two, 1,632-horsepower diesel engines she was able to cruise at 14 knots with the assumption that both engines were in solid working order, which was not the case on September 26, 2002. Between September 13, 2001, and September 10, 2002, the *Joola* had actually been out of commission, as she had undergone mechanical damage repair and replacement of the portside engine. The ferry had only resumed operation for two weeks when it capsized and sank off the coast of The Gambia. The *Joola* had been rushed back into service for the launch of the school year and for its "sea trial" from Dakar to Ziguinchor. One new engine had been installed, but there was no time for scheduled maintenance on the ship's second engine. On September 10, after the Dakar-Ziguinchor sea trial, the government declared the *Joola* safe to sail (Becker 2012). But with only one operating engine and no major overhaul on the second one in five years, the *Joola*'s speed was more realistically ten knots.

Many critics, including my colleagues in Ziguinchor, pointed out that the hasty move to return the *Joola* to sea coincided with Senegalese elections, the war in Casamance, and a need to restore the optics of national unity in the south. As one shopkeeper in Ziguinchor noted:

> The government did not put the *Joola* back in operation because people in Ziguinchor were pressuring them to do so. They did it for political reasons, because elections were coming up. It wasn't about us, it was about them. They wanted to give the illusion that all was well in the country, but clearly it was not. They put that ship back into operation too soon, and there were terrible consequences.[12]

Invoking such responsibility for the ferry wreck—while not the central focus of this narrative—has been widely debated at the local, regional, and state level. *Joola*'s seagoing capabilities and service record have been repeatedly questioned as evidenced by the 216-page High Court of Every report (Thomas and Tricot 2008). Simon Boissy, who managed the *Joola* company from 1990 to 1995, admitted that the shipwreck disaster was the result of poor maintenance. A native of Casamance who lost family members in the event, Boissy stated that the ship was designed to last thirty to forty years, yet by 2002, after only twelve years in operation, she was found to be in a poor condition (Technical Commission of Inquiry 2002). Federal officials, including those in the military, would later dispute Boissy's notion, framing the wreck as a function of weather. Locals I spoke with believed that the government used weather as a red herring in order to divert attention away from the state's own culpability and need to politicize the ship's return to sea.

Unlike many passenger ferries in Africa, the *Joola* was operated by the Senegalese military since its inception. The boat provided a means to unify the north and south by way of an affordable maritime transport, but she also served as a practical mechanism to help transport the military back and forth between the two sectors of the country in order to oversee the Casamance conflict (Auvray 2012; Becker 2012). But because this merchant ship was controlled by the military, boarding and cargo loading of the *Joola* followed military rules while flouting international standards and maritime regulations. *Joola* passengers therefore faced the worst of both worlds; they were traveling on a merchant ship operated by a military apparatus that was unwilling to apply standard basic maritime protocols. Commander (Captain) Issa Diarra, who perished on the ship that night, may be absolved of some responsibility given that he was forced to operate within the confines of a military structure, but his willingness to sail into "suicidal conditions" on September 26 while understanding the mechanical risks is inexcusable (Thomas and Tricot 2008, 112).

The official investigative report by Thomas and Tricot (2008, 110–111) confirms that no one on the *Joola* was monitoring the stability of the vessel during its operation and this was also the case on September 26.The role of a vessel's energy officer is to verify and adjust the stability of cargo but the *Joola* had no such officer onboard the ferry to monitor loading and ballast. As a result the loading process did not comply with the design calculations of the ferry. Some of the cargo loading of fish, palm oil, mangoes, and other produce were properly placed in the cargo holds, but the loading of passengers and vehicles was another matter (Wiley 2012, 45). It is difficult to argue that there were any solid procedures in place to ensure compliance with the ship's loading and stability requirements. Moreover, there is no proof that the crew onboard the *Joola* was adequately trained in areas of maritime protocol. We know that children under the age of five were not required to hold tickets and those of limited means were allowed to ride for free. The weight of these ghost passengers and their cargo was never calculated during the boarding process, nor was adequate attention paid to the high center of gravity as passengers flocked to the upper decks.[13]

PUBLIC SPACE AND TERANGA

This problem of overcrowding is not unique to Senegal and is tangentially related to the notion of *teranga*. Teranga means "enduring hospitality" and can include the way that Senegalese provide a welcoming space for those who need to board a bus, a taxi, a pirogue, or a ship despite the available spots. In the coastal village of Somone, for example, taxi drivers transporting someone to market will stop, without questions asked, to pick up an elderly person and transport that person free of charge. This taxi example is true even if the driver has a full vehicle; all are welcome, whether or not space permits. *Joola* ship workers behaved in accordance with these same *teranga* principles throughout the ship's lifecycle. Oftentimes ship workers would let people board if they were missing their Senegalese ID or even if they failed to purchase a ticket at all (Becker and Diallo 2012). While this point is not meant to implicate *teranga* as the key cause of the crisis, in the case of the *Joola* it is clear that Senegalese were used to being flexible when it comes to available space. Formal rules were and are often ignored in the name of hospitality. This openness certainly precipitated the disaster, though it was not the major cause of the ferry wreck. Senegalese were accustomed to overcrowded transport routes and tended to "roll with it" and leave safety in the hands of Allah.

Yet it is important to note that overcrowding was not only a function of ship workers' hospitality and flexibility with regard to those of limited financial means. Many who boarded the ship that day, like on other ordinary days, had bribed ship workers so they would be given a free pass. Just as *teranga* is part of Senegalese culture, graft is also embedded in the country's economic systems. It is a painful reality, but the truth is that those ticket takers who accepted money in exchange for looking the other way upon passage remain partly culpable for the disaster. Likewise, while passengers, native to Senegal, are comfortable with crowding, many were in fact unsettled about the ship's visibly heavy load that day. In interviews I was told of how in a twist of fate some potential passengers had opted to disembark the ship, deciding to wait another day to travel. Many passengers were scared because while it was normal for the *Joola* to be beyond its carrying capacity, it was atypical for the ferry to be overcrowded by 1,000 persons. But in September of 2002, historical and economic forces were at work that shaped this stark reality. One cause for overcrowding was the fact that residents of Ziguinchor had had their transport systems sidelined while the *Joola* was out of operation for mechanical repairs. Thus there was a sense of playing catch-up once ferry services resumed, particularly for those rushing back to school in Dakar for classes, or for vendors needing to get products to markets in the north. The disaster was, as President Wade remarked, an accumulation of errors (IFRC 2002). While I agree with the president's assessment, the wreck was also an accumulation of everyday cultural practices leading up to the perfect storm, driven by a set of norms that had become widely accepted over time. Ship workers let children pass by, without counting how many actually boarded, and similarly, they gave permission to their brothers and sisters to hop on the ship, sometimes in exchange for a few CFA in payment. One consequence of this process is that the government will never have an adequate account of just how many perished that day, which is especially true for large numbers of the nonticketed youth (Becker 2012). Dr. Charles Becker (2019 interview) took on the arduous task of gathering information on *Joola* victims in order to create the list found in the Appendix. This data project was community driven rather than a function of good government recordkeeping.

THE CULTURE OF GRAFT

While Senegal has made great strides in terms of peace and conflict, research by Transparency International shows that corruption remains a major concern in a country where "resources are poorly managed and government spending remains opaque" (Transparency International 2012). In a TI Global Corruption Report, some 88 percent of Senegalese noted that corruption had become worse in the past three years (Transparency International 2015). Note that this timeframe for ongoing corruption follows in the wake of the *Joola* disaster, suggesting that government misconduct continued even after the disaster. Incoming president Macky Sall promised new jobs, better education, and, most relevant for the *Joola* narrative, *accountability* from political leaders (Transparency International 2015). Sall vowed to tackle corruption in Senegal, and the Transparency International's partner agency "Forum Civil" has worked to gather real life stories of how corruption is affecting the lives and livelihoods of ordinary cities—including access to education, health, water, roads—encouraging them to demand accountability. As a result there have been public rallies from Bambilor, Kaffrine, Saint Louis, and Dakar—as part of the national "No Impunity" campaign. The message for Senegalese people is that they need to work with local governments to identify corruption problems and solutions, that they should not accept bribes for services that should be free, and that corruption should not be perceived as a way of life (Transparency International 2015). Sadly, the latest BP Oil Controversy sheds light on the rampant corruption that still exists at the state level more than two decades after the *Joola* disaster.[14]

SEPTEMBER 26, 2002: SENEGAL'S 9-11

Details relating to the day of the shipwreck were recounted to me by more than four dozen individuals during my time conducting research in Senegal. These particulars have been confirmed in the *Naufrage du Transborder Senegalis*, the official report of the *Joola* tribunal (Tricot and Thomas 2008). In November 2018, Mustapha, a fifty-two-year-old resident of Casamance, described to me how September 26 began as just another ordinary Thursday in Ziguinchor, with a few important exceptions.[15] In Islam Thursday is an important day in which to undertake a journey, as the holy Prophet Muhammed chose this night to travel. The *Joola*'s departure had always been a part of the regular rhythms of Casamance daily life, where tourists, diplomats, and businessmen staying at the Hotel Kadiandoumagne—a small, colonial style hotel in downtown Ziguinchor staffed mostly by Djola people—would watch as the ferry prepared to disembark with a new round of passengers headed back to Dakar. On September 26, the ticket office had opened at 0830, and because the ship had only recently resumed operations, by 1000, roughly 2,000 people already filled the portside dock (this book uses military time). The standard twice a week Ziguinchor-Dakar rotation had been reduced to one scheduled journey in the wake of ship repairs, so locals were now eager to grab any seat available en route to the north given the limited space. The initial weather forecast broadcast to Ziguinchor's radio station called for scattered storms, not unusual for this time of year, and the Technical Commission Inquiry reported Ziguinchor's conditions as "calm wind, good visibility, cloudy sky" (Republic of Senegal Investigation Report 2002, 62). Mustapha, like many others who were present to witness that day, reported one noteworthy difference: that by 1230 the ship was severely overcrowded, more than was usual, and she was tilting to port in response.

After disembarking Ziguinchor's bustling harbor at 1300 the *Joola* traveled three hours down the Casamance River, affording passengers an opportunity to admire the serpentine *bolongs* or tributaries common in the south, most lined with enormous kapok trees, some of which are 700 years old and 70 meters high (Figure 4.2). This verdant area of Senegal, sparsely populated once you leave the metropolitan area of Ziguinchor, is replete with some of the most diverse cultural and environmental landscapes in West Africa. Many Casamance people live scattered throughout these remote bushlands, and so the ferry journey provided a rare opportunity to visualize a larger world outside their village. While the Senegalese Ministry of Transportation often hired musicians to pass the time onboard, many passengers—whether Djola, Mandinke, Sereer, Pulaar—instead opted to crowd on deck to admire the geographic sideshow of tropical Casamance. The ship journeyed 45 miles before arriving at Carabane Island at 1630 without any noteworthy incidents.

As the only scheduled stop between Ziguinchor and Dakar, Carabane Island had additional passengers who had been waiting to board what was an already overcrowded ship.

According to witnesses, nearly 12 *pirogues* with 30–40 passengers each were preparing to load on the starboard side of the vessel (Wiley 2012). Not only passengers, but cargo was also being readied for the journey. Dried fish, shrimp, rice, and tropical produce like mangoes from Casamance were destined for the large markets of Dakar. As pointed out in chapter 2, the south was and continues to be natural-resource–rich relative to the northern part of the country. The Ziguinchor passengers helped assist people arriving via the small *pirogues*, while others expressed shock that more passengers were being loaded onto an already overcrowded ship. From the top deck Ousmane Keita recalled how "it was shocking to see the hundreds of passengers that were waiting to board from tiny pirogues. It was so hot that so many of us went to the top deck to witness what was unfolding."[16] In Carabane at the time, there was no formal point of entry or exit for passengers, and so *pirogues* served as transfer vessels for the anchored *Joola* just offshore. While survivors Ben Bechir Badji and Ousmane Keita estimate that hundreds more boarded the ship, other statistics indicate that 185 additional passengers were loaded at Carabane that day (Auvray 2012). Though no one is certain of the precise numbers due to inconsistent recordkeeping, the ship was clearly beyond capacity and existing passengers were worried (Barton 2017). As a result some passengers opted to disembark the *Joola*, some of whom departed at the prospect of an uncomfortable night of sleeping on a crowded vessel, especially for those in third class and no berth in which to sleep.

With a total of 57 square kilometers, Carabane (or Karabane) is the last major island and village of the same name in the mouth of the Casamance River in southwestern Senegal. Glimpsed from satellite image above Carabane Island tucks into Casamance like the last puzzle piece in the country's meandering south. From a geographical perspective the island is simply a marvel, replete with reptiles, coastal manatees and crocodiles, collared vultures, and anhingas as black as coal. The island is easily accessed by a two to three hour boat ride from Ziguinchor; a motorized *pirogue* trip from Carabane to Elinkine, the nearest village, takes about thirty minutes. Historically, Carabane was a Jola inhabited space until the arrival of the Portuguese from the sixteenth century onward. Since World War II, the village has had small population densities due to drought, the Casamance conflict, and ultimately, the sinking of the *Joola*. Home to Basse Casamance National Park and Kilassaye Avifaunal Reserve, the island supports fruit crops such as mangoes and oranges and an abundant artisanal fishing industry. As an important historical site of Senegal, residents here have been reluctant to open themselves up to the burgeoning ecotourism industry. While the MFDC ceasefire of 2004

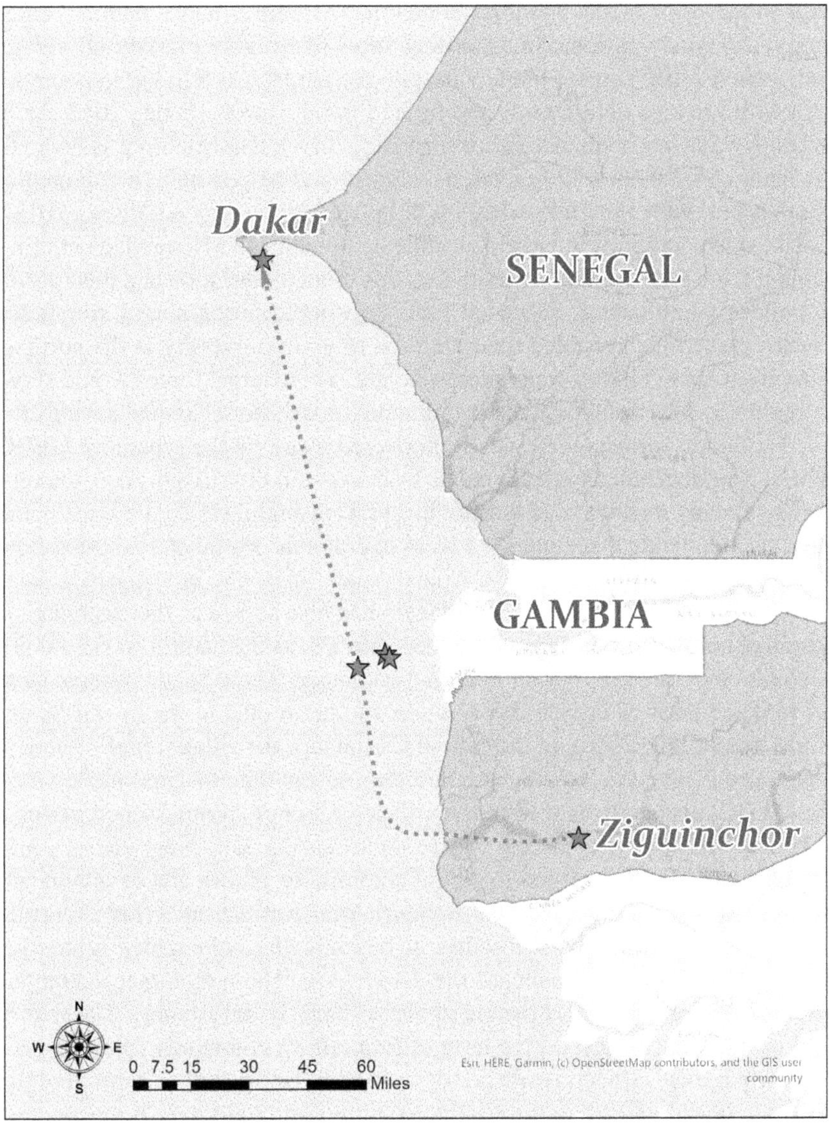

Figure 4.2 Map of the *Joola*'s Route. Cartographer: Dr. Jieun Lee.

brought peace to the island, the sinking of the *Joola* claimed many inhabitants of Carabane and weakened its ability to engage in trade for several years after the event (Barton 2017).

We know that of the passengers who boarded at Carabane that day a large majority of them hailed from the nearby village of Mlomp. A total of fifty-four of those passengers, primarily students, lived in Mlomp, just off a dirt road in the heart of the Casamance conflict. Mlomp is an exceptional community for many reasons. In a startling show of resilience this small village had achieved 100 percent literacy despite the hardships of living in a remote area with brick or mud houses and limited job prospects (Wiley 2012, 217). In some ways this resilience is not surprising, given that Mlomp was home to the team of students who funded, developed, and helped build a community youth center with electricity, the first of its kind in the region (Auvray 2012).

Like most activities in larger Casamance, students in Mlomp had enduring support from their community both in terms of their education and their desire to build a youth center. Extended families worked in the nearby rice fields to earn cash, which enabled their children to go to university in the north as what we would call first generation scholars. Yet external forces would serve to challenge Mlomp's long-term community goals. For example, savings for the community project were nearly hoovered away by the separatist MFDC rebels, who demanded cash payments from those in the village. Also, the rebels had already mined a road in Mlomp and were angling to do the same in the adjacent rice fields that supported local livelihoods. Fortunately, at that time the nephew of Mlomp's chief—a renaissance man who envisioned a solar powered electrical system in the village—had also served as the custodian of the sacred woods. Such spiritual spaces with ties to the natural world are not uncommon in Senegal, particularly in Casamance. These belief systems were set in place prior to French colonization and are rooted in traditional Sereer worldviews (Sene 2016). In the case of Casamance the village chief's nephew threatened a curse on the rebels should they co-opt the villagers' money, and the MFDC rebels retreated from the village in fear of spiritual repercussions.

It was an inspiring success story, underscoring how Senegalese youth thwarted the MFDC's efforts to curtail community growth and development. What is most tragic is that these students, who worked the fields alongside their parents, eventually succeeded in building the community center yet would ultimately perish onboard the *Joola* ferry. The notion that a community so small could lose a demographic so large is devastating. The reality of this loss extends beyond the term of *brain drain* commonly found in geographic parlance. The consequence of losing fifty-four youth, a symbol of the village's hopes and aspirations, is something that could only be reconciled through the region's faith-based traditions. Resilience did not come easy after the shipwreck, but this was especially true for disproportionately impacted communities like Mlomp.

While Mlomp is geographically closer in proximity to Ziguinchor than to Carabane, many students preferred to board the *Joola* in Carabane because it was considered safer. In order to reach Carabane students had to first travel to the village of Elinkine, which is located on a tributary of the Casamance River, and from there they hired a *pirogue* to reach the island in time for the journey. Neither direction was trivial, but traveling from Mlomp to Ziguinchor involved using a poorly maintained road. Not only that, but the transport route also served as a conduit that was frequently attacked by MFDC rebels and thus regularly avoided by locals. The deeper one digs into the *Joola* story, it becomes clear that military politics both enabled and constrained the everyday geographies of local Senegalese people. Everyday economics and ongoing separatist conflicts determined how people moved across time and space as well as why.

Yet not everyone boarding the ferry in Carabane hailed from Mlomp; others lived in surrounding villages, including *Joola* survivor Ousmane Keita. Ousmane described to me how he traveled with the chief of his village of Guet Ndar; Mr. Keita was headed to the University of Dakar as a freshman interested in studying geography. When we met, we connected given our shared interests in the geographic discipline.[17] Kind and soft spoken, Ousmane was older than many first-year college students in the United States at the time of the 2002 shipwreck. However, his status as a nontraditional student was not unusual for Senegalese people. Many others like Ousmane needed to first meet family commitments for their families before pursuing a degree. In numerous cases, extended families were also tasked with raising funds for university and the associated costs of moving north since there were no postsecondary options in their backyard. Yet despite financial struggles in the south, the Casamance region boasted the highest percentage of students in the country going to university (Wiley 2012). Thus while families in Casamance have long struggled with the separatist war, underfunded secondary school systems, and a lack of higher educational institutions, their highest priority was still placed upon education. And yet, students from all over Casamance—not just coastal Ziguinchor and Carabane Island but villages in the bush country as well—needed to board the *Joola* in pursuit of education not found in their own backyard.

THE STORM OF SEPTEMBER 26, 2002

It was within this geographical setting that the *Joola* departed Carabane Island and continued northward on its standard maritime route to Dakar. Several observers snapped photos of the boat listing to port as it left the small island community. We know that the last call from the ferry staff was broadcast to a maritime security center in Dakar at 2200 and "good travel conditions" were reported by the *Joola* crew (Becker 2012).[18] Yet this conflicts with the weather reports coming out of Météo-France (Toulouse) via Inmarsat, which were now predicting violent gusts of 15 to 25 knots in the eastern Atlantic region near Senegal.[19] Likewise, the Commission Report

Technical Survey (2002, 48) forecast confirmed "heavy rains, thunderstorms, very strong winds of 20-40 knots for 15 to 30 minutes" for the region into which the ship sailed as it passed the Gambian coast. Unfortunately, this severe weather forecast was distributed to Port Autonome de Dakar but was not broadcast by the Dakar-Radio maritime radio station. To access this key information, the *Joola* would have had to contact Dakar or the nearest weather station, but Mr. Amadou Maiga, the meteorologist for Ziguinchor, confirmed that no member of the *Joola* crew had reached out to his office to acquire an updated forecast.[20]

This type of tropical storm, which moves rapidly onto the coast of Senegambia, was actually not uncommon for the month of September in West Africa. So it was shortly before midnight that the *Joola* sailed into such a storm, the rough seas and high winds contributing to the ship's capsizing off the Gambian coast. At the helm, Commander Issa Diarra would have had past experience with these conditions; according to the meteorological record the sea had been building and by 2100 the waves averaged around 2.5 meters, with winds gusting at 30-40 knots (Tricot and Thomas 2008, 146). As a result of the rough seas and high winds, the ferry overturned, throwing passengers and cargo into the sea, all within the space of five minutes. French survivor Patrice Auvray reported that the event happened so fast, that there was no time to reach the lifejackets or rafts because they were tightly secured (Auvray 2012, 19). The Expertise Report (Thomas and Tricot 2008, 66) confirms Auvray's firsthand observations; there was clearly an abundance of rescue machinery onboard and though insufficient for the number of passengers it was not accessible in part because of the speed of the wreck.[21] The *Joola* actually contained two covered motorboats under davits with ninety-person capacities per boat; twenty-two inflatable life rafts that accommodated twenty-five people each; two zodiac rescue boats; thirteen life buoys; and approximately 667 lifejackets.[22] But at the time of the sinking most of this equipment had no hydrostatic release systems in place and they were therefore permanently seized on deck. Patrice Auvray wrote in his memoir:

> I did not notice life jackets in the cabin, where can they be found? I can see the crates of life jackets and the life rafts strapped to the deck. . . . Impossible that in such a short time someone will be able to free them. We should already reach them![23]

In the words of Thierno, a shopkeeper friend in Ziguinchor: "The *Joola*'s sinking was only the length of a song."[24] It is clear that in less two minutes the ship was "laid on its port side," where it sat for a few minutes before water began to rush through the portholes of the main deck and the stern's door (Thomas and Tricot 2008, 145). When the incline exceeded twenty degrees, an electrical power failure resulted, plunging the ship into darkness, and as Ousmane Keita and other survivors described, this problem added to passenger panic (ibid, 146). When the ship finally sank at 2255, it was located at 13° 12' 8" North, 17° 5' 6" West, approximately 16.8 miles from Point Saniang in The Gambia.[25]

The wreck drifted for more than three days toward the East-Northeast, eventually penetrating into Gambian territorial waters and hitting the sea floor at 1400 on September 30 (13° 15' 3" North, 16° 59' 3" West). Three days after the sinking divers filmed the wreck. It had drifted toward the coast until it rested at 20 meters depth; the bent masts of the *Joola* came in contact with sand, representing the only visible damage to the ferry (Thomas and Tricot 2008, 146). By October 17, 2002, the wreck lay on its port side, heading east, located at 13° 14.9 North and 16° 58.4 West in Gambian waters. At least 1,863 people perished, although the exact number will never be known due to the large number of unticketed passengers onboard. Among the dead were 1,201 men (61.5 percent) and 682 women (34.9 percent), although the gender of seventy victims is still unknown (Barton 2017; Becker 2012). The *Joola* manifest included passengers from at least eleven countries: Cameroon, Guinea, Ghana, Nigeria, France, Spain, Norway, Belgium, Lebanon, Switzerland, and the Netherlands (Barton 2017; Becker 2012). Not one American was onboard.

Many officials would later claim that these critical weather conditions not only were factors in the capsizing but also shaped the passengers' ability to survive in the rough Atlantic waters. We know that the tropical climate of Senegal involves a long dry season that runs from mid-October to mid-June in the north, and early November to mid-May in the south (Liberto 2015). The rainy season, in contrast, runs from late June to early October up north, and late May to late October in the south. This phenomenon is due largely to the African monsoon, which upon its arrival in Casamance brings up to 60 inches of rain to the region, compared to the north and center of Senegal, which is part of the Sahel and receives only 23.5 inches per year (NOAA 2019). In terms of associated storm systems, scientists know well the way in which easterly winds carry tropical storms away from the African continent during this season, which was also the case in 2002. According to the NOAA, however, September 2002 was the "most active month of any year on record with the formation of eight named storms" (NOAA 2019). Four of these storms went on to become hurricanes in the Americas (Liberto 2015, 1–2). Typically, Atlantic hurricanes do not often impact West Africa or the Cape Verde islands directly, but rather they begin their lives as atmospheric disturbances just offshore Senegambia, the Cape Verde islands, or Canary Island, dropping damage as they head further west into the Atlantic (Lagos 2002). Minor West African disturbances in due course can become America's climatic nightmares in the form of hurricanes. But in the face of climate change, atmospheric disturbances are now transforming themselves into hurricanes east of 30 West (Pasche et al. 2020).

Traditionally, it takes storms time to fully develop once they leave the West African coastline, whereby they intensify to hurricane level after several days moving over warm ocean waters. But in 2015, Fred evolved into a tropical storm at 18.9 degrees west longitude and a hurricane by the time it traveled to 22.9 degrees. In terms of history, only eleven hurricanes have formed east of 30 degrees west, let alone as far east as 20 degrees west (Pasche et al. 2004). Fred set its sights on the Cape Verde islands, just offshore Senegal, which prompted the first hurricane warning since 1851 (NCEI 2002). In 2018 I witnessed a storm decimate a hand-built restaurant on the beach of Somone, wiping out a Senegalese man's business in hours. Cape Verde and Senegal are both accustomed to these violent storms that originate in the region, but as noted in our discussion of Saint Louis's evolving coastline, climate change is altering patterns in unprecedented ways, with Africa being the most likely to be impacted (World Food Programme 2011).

In the wake of the disaster, Senegalese officials including the navy blamed the *Joola* wreck on weather while downplaying other forces that were more human in nature (Senghor 2004). Just days after the disaster Navy Commander Ouseynou Combo stated: "There was no problem of weight or overloading of a nature that would cause this situation" (Rothe et al. 2006, 1). Prime Minister Mame Madior Boye added: "The boat underwent a technical inspection in September, and I want to say, according to the information we have at this time, it was the rain and the wind, natural phenomenon following a storm, and the boat capsized" (Lombard 2003, 1). On September 28, 2002, he continued: "For the moment, the condition of the boat is not suspected" as a cause of the accident (Gounongbe 2003, 1). Frigate Captain El H. Habibou Faye was even more certain: "Concerning the conditions, the circumstances of the accident, it was *completely* due to bad weather, technical problems cannot be blamed" (CNN 2002). Adjectives and emphases were chosen carefully to underscore the climatic rather than human culprits. He added: "The crew of this ship was hand-picked by the national marines. Commander Issa Diarra had extensive experience, 21 years of service, and he also was hand-picked to command this boat. The National Marines would not pick just anyone to serve on this ship" (Gounongbe 2003; Wiley 2012).

Not only did the navy downplay the role of overcrowding and the faulty engine repair, but they elevated the role of West African storm systems in terms of the ship's demise. In the wreck's aftermath the navy cited a fierce, brief gale that had blown into the eastern Atlantic into which the captain had directly sailed (Barton 2017). Such a notion of "acts of nature" helped legitimize the government's poor handling of the crisis, the anemic rescue operation, and the root structural causes of the disaster (CNN 2002). It was also interesting that the storm of September 26 was likened to the *Titanic*'s iceberg, yet framed by the government as an unavoidable tragedy beyond human control. Storms and icebergs may be avoided given accurate maritime intelligence and proper communication technologies. Many of those I interviewed between 2016–2018 mentioned poor weather on the night of the disaster, explaining how it must have challenged the rescue operation. They discussed the difficulties of trying to survive for several hours before first responders arrived. However, not one individual mentioned the storm as the *root* cause of what would become Senegal's worst humanitarian disaster. Weather was seen as a complicating factor in what was an avoidable human rather than natural disaster.

Senegalese officials were not the only ones to cite the role of the storm in the *Joola*'s demise. Local journalist Diadie Ba described to the media how "The weather conditions were very, very bad. We know it was very windy and there was a big storm" (Thomas and Tricot 2008, 5). Yet it is the reports of the survivors who should carry the most weight in terms of what really happened that night and the factors that precipitated the event. They were, after all, right there. One of the survivors, Moussa Ndong, told the Associated Press by telephone in his hospital in The Gambia: "The boat went down so fast. It was so unbelievable—in just three minutes, the boat went down. It was horrible, because we heard people screaming underneath" (Gounongbe 2003, 6). Ndong described how the storm brewed quickly and the wind gained speed and as the boat began tipping to one side, people jumped off the deck as water rushed into the main cabin. When the lights went out, the passengers began to panic, and many of those stayed on top of the capsized vessel for several hours before the fishing boats arrived on the scene. Survivor Moula Badji described similar environmental conditions of strong winds blowing water onto the deck, of heavy rain pouring more water into the ship before the boat began to topple. "Everything happened so quickly. The boat overturned in less than five minutes." Cheik Niang remembered a flood of water as the *Joola* turned upside down: "The sea was really choppy, rough, and cold." While many passengers clung to the capsized boat, others had to struggle to stay afloat in the ocean. "After the boat capsized, we were thrown about by the wind and huge waves until about four in the morning," said survivor Ben Badji (Technical Commission of Inquiry Investigation Report 2002). Many floated close to the boat and clung to one side until fishing boats arrived at the scene (Wiley 2012).

When the ship left Carabane at 1800, an onshore tourist who filmed the ship leaving the island famously reported how she could be seen listing to port.[26] Other interviewees would make the same observation. The weather forecast for that night reported: "scattered thunderstorms, winds from west-northwest, with gusts of 35-45 knots and two meter seas." Given the movement of pressure belts as they relate to the African Monsoons in September, the forecast was not unique, and the *Joola* had encountered storms before. The only unusual aspect of the weather report was that winds were expected to originate from a direction that would exacerbate the ship's listing to port. Though none of the survivors on the deck of the *Joola* witnessed a big wave, it would have been difficult to spot given limited visibility during the 2300 time frame (Becker 2012). But the high winds listing to port combined with a large wave could have caused a domino effect as passengers, fuel tanks, and unsecured cargo shifted to one side to a point of no return, rendering the *Joola* unable to right itself once it began to keel. The boat, with its shallow-draft and flat-bottomed hull designed for riverways, would begin to fill with water, via open windows and then down into the cargo and garage sectors. As many survivors noted, there was no time to recover when the *Joola* capsized. "It all happened so fast, and nearly 20 miles from the coast, there were no ships there to witness the event as the hull overturned dumping 1800 passengers into the Atlantic" (Mane 2018).

WOMEN AND CHILDREN FIRST

The topic of gender in Africa is a complex one, and because it is not the central focus of this *Joola* book it cannot be given adequate treatment. Yet it is crucial to explore gender within the context of the shipwreck, because when the *Joola* sank off The Gambia, *only one* of the sixty-four survivors was female. Despite the presence of nearly 600 female passengers onboard, Mariama Diouf, who was pregnant, was the sole female passenger to have endured. At the time of the wreck, Mariama was a thirty-seven-year-old mother of six children, five boys and one girl, ranging in age from five to seventeen. Ms. Diouf had first traveled from Dakar to Casamance via bush taxi in 2002 in order to visit her family compound of thirty-six people in Ziguinchor. She had planned to purchase goods of local Casamance handicrafts to sell for a profit upon returning to Dakar. Raised among a family of fishermen who lived not far from the port of Ziguinchor, Mariama was a product of Qur'anic religious schools, which provided free education for the first six years. Since the French school system involved the purchase of books, many poor families could not afford this standard option and so they chose Qur'anic institutions instead. Ms. Diouf, a strong swimmer given her family's fishing roots, had purchased a third-class ticket for the *Joola* with a return date of September 26.

The possible reasons as to why Mariama was the only woman to have survived the wreck cannot be overlooked, as the reality defies statistical odds.[27]

A female survival rate of one forces us to ask deeper questions about (1) whether the maritime mantra of "Women and Children First" was instituted onboard the *Joola*; (2) the role of women in Senegal including their experiences in oceanic environments; and (3) the relationship between gender and class in terms of where women were situated on the boat when it capsized. Each of these variables is explored in turn, keeping in mind the paucity of information available since the boat sank so quickly. Much of this discussion is speculation based on interview data but is nevertheless worthy of investigation. And while this section does its best to understand events that correlate with the low female survival rate, I am cautious not to imply causation. The truth is that the process was complicated and is related to numerous cultural factors, including the way in which society and space are gendered both in Senegal generally and on the boat in particular.

In presenting this work on the *Joola*, audiences frequently asked about the gender breakdown of passengers and survivors. At a talk in Boston University (BU), a professor inquired as to whether the "Women and Children First" (WACF) mantra was employed on the ship, as the audience was startled to discover that Mariama Diouf was the sole female survivor despite the presence of many female passengers in all age brackets.[28] Here in the West we believe this WACF motto to be fact; films like *Titanic* have helped us to internalize the narrative that when a ship goes down, women and children have priority boarding on whatever rescue vessels are available. In truth, "Women and Children First"—or to a lesser extent, the Birkenhead Drill—is a code of conduct which suggests that the lives of women and children were to be saved first in life-threatening situations, but more generally it was applied to the case of abandoning ship when lifeboats were scarce. While WACF has no foundation in maritime law, it is a commonly perceived practice, dating back to 1840 when after a lightning strike a fire broke out onboard the American *Poland* as it was en route from New York to Le Havre (Biel 2012). According to passenger J. H. Buckingham:

> The captain said that he had little doubt that the ship was on fire, and that we must endeavor to get at it. One suggestion that we might be obliged to take to the boats, it was immediately remarked by one of our French passengers, and responded to by others—Let us take care of the women and children first.

A similar event occurred onboard the HMS *Birkenhead*, a Royal Navy troopship that required evacuation in 1852. Yet the most famous application of the WACF drill is that of the RMS *Titanic* in 1912 (Biel 2012). Many argue that as a consequence of this motto in action, some 74 percent of women and 52 percent of children onboard the legendary ship were saved, while only 20 percent of men survived (Biel 2012). While modern-day disaster evacuation experts now argue that we must help "the most vulnerable to leave the scene first, likely to be injured, elderly and young children," extant programs such as the Boy Scouts of America's Sea Scouts still practice "Women and Children First" as their motto of the sea (BSOA 2019).

Recent research shows that the WACF motto is less accurate than statistics indicate. For example, while the *Titanic*'s high female survival rates were a product of the protocol, these statistics cannot be separated from class and privilege. For example, while 20 percent of *Titanic*'s male passengers survived—when disaggregated, we learn that 32 percent of men in first class endured, whereas only 8 percent in second class, and 13 percent of those in third class (steerage) would live. The statistics are more startling for women. Some 97 percent of women in first class survived, whereas only 49 percent in steerage received access to the lifeboats. The statistics for children are more mixed with an overall survival rate of 52 percent for all classes. Yet the evidence is clear: youth who were in first, second and third classes had survival rates of 86 percent, 100 percent and 31 percent, respectively (Biel 2012). To be clear, the economic status of passengers and their physical positions onboard the ship featured prominently into whether those individuals would survive or not. Therefore while we have in some ways internalized the motto of the seas, we have failed to give it proper treatment from the perspective of class, both in the West and in Africa.

The answer to the question of whether the mantra of WACF was implemented onboard the *Joola* is fairly clear. It was not. Much of this has to do with Senegal's lack of informal or formal policy in terms of the protocol of maritime disasters. Also, unlike *Titanic*, access to lifeboats was insufficient and inaccessible.[29] Because it took only minutes for the boat to capsize, there was simply not enough time to even reach the life jackets in order to abandon the ship into the Atlantic. Nevertheless, the way in which the disaster disproportionately impacted women was also a function of their physical position onboard the vessel. This notion does not necessarily apply to female tourists, who were traveling for leisure rather than business, and free to roam the boat as they saw fit. Unlike the market sellers or mothers of children, international tourists were not constrained to certain sectors of the boat for the duration of the journey. They had, what geographers often call, a certain degree of spatial freedom that was removed from the local culture and its long-standing gender norms. But most Senegalese mothers were tending to their children when the ship went down at night, in contrast to most men who were on the deck listening to music or sharing food with friends. In the words of nineteen-year-old Astou, who I interviewed in Ziguinchor in her apartment building:

> I lost my mother onboard the ship that night. She was the rock of my family. She was the foundation. She held everything together, and when she passed

> away, my father went crazy, he went mad. I am young but it's my job to take care of my brothers and sisters now and I get some help from my uncle but it is still difficult.[30]

Some sixteen years after the event, Astou is embodying the same gender norms and expectations of her mother, perpetuating existing societal roles reinforced by the very tragedy.

Thus one plausible argument is that women onboard the *Joola* had a different set of societal expectations which may have hindered their escape. Many would have had to rise early the next morning to tend to their children, or in the case of mango or market sellers, they would need to sleep early so they could wake as soon as the ship reached Dakar. The majority of these women and their children were trapped inside the hull when the ship sailed into a storm near midnight and capsized, with little opportunity for easy escape (NOAA 2020).

But not all women in Senegal fit into these traditional gender roles at that time, which further complicates this story of survivorship. While the mango sellers remained on the first level of the *Joola* with their produce, and other women remained in the berths with their children, there was also a group of women who were on the top deck that night. Some were in this subset of women heading back to the university in Dakar like their male contemporaries (or in some cases, working as prostitutes). One colleague told me, "The *Joola* had a stigma, in some ways, as a party boat. Both young women and men would stay up late on the overnight journey, eating food, listening to music, smoking, in some cases drinking." Many parents did not want their children to take the boat because of this problem. They worried less about their drowning in the Atlantic and more about the mischief that young people might get into at night. But still, some young women, despite Senegalese cultural norms, used the *Joola* as a transport route to make their way north. They stayed up late, which means many were not trapped in the berths like the other older women when the ship capsized. As one young male fisherman from the coastal village of Somone noted: "Many women were aboard the *Joola* that night, just like the other men, so when the ship capsized, they also could have had the opportunity to survive. But for some reason they didn't. All I can say is that, as a fisherman, where you are on a boat matters. Especially if things happen fast."

THE SEA AS A GENDERED SPACE

While it is difficult to speculate what may have occurred had women not been trapped in the hull tending to family, some people I spoke with—particularly fishermen in Somone or Ngaparou—argued that one of the key problems of the crisis was that both men and women lacked basic swim skills. The "Geographies of the Sea" chapter has outlined the conflicted relationship that many Senegalese have with oceanic and riverine spaces. While Senegal is a fishing-based state, the economy of which forms part of the national identity, beyond those fisherfolk working on *pirogues* the vast majority of people do not swim. Not only do young people grow up without reliable access to swimming pools, but there is a widespread fear of water bodies given the presence of crocodiles and hippos. This renders swimming more of a functional or economic activity than a recreational one. While on the night of the disaster there also was nowhere *to* swim, given the lack of adequate lifeboats and the state's failure to execute a rescue mission, many Senegalese still blamed reduced swimming skills as part of the problem. Locals would frequently explain to me that women in particular did not have experience with sports, particularly in the area of swimming. As thirty-four-year-old Mustapha of Ziguinchor noted:

> Most of us believe that it is Allah's will, what happened. It is the only way to explain it. But the fact that only one woman survived was odd. That she was pregnant was an act of God, but it is still odd. Most women were trapped, we have heard. But I also think the problem is that in our culture women do not do sports. It is not accepted culturally, it is not common. It is changing, but few women grow up running and very few learn how to swim. You can see young girls playing in the waves of the river, but that is all you see.

Mustapha's words were made real during my fieldwork in Senegal. Whether on the beaches, outdoor gyms, soccer field, or popular jogging routes of Dakar, it was far more common to see men "doing sport" than women. According to forty-year-old Marie Ly, mother of two, who lives adjacent to Ngor Beach in Dakar, "I don't know how to swim. I am too busy cooking and too tired from cleaning to take care of the kids or do sport."[31] Marie's response is not surprising, given the traditional role of women in Senegal. Most agree that the unfair burden of domestic work still falls upon women, even in modern urban regions. Moreover, many women have been left as heads of households in cases where their husbands have fled to Europe in search of work, left with little free time to pursue leisure activities.

It is true that women in Senegal face several disparities in terms of their social status, though this is slowly shifting over time, particularly in urban sectors such as Dakar. Women have high rates of illiteracy across the country, in part because of the requirements embedded in a culture where women must first care for family and children. As such women comprise only 10 percent of the formal labor force, though the informal sector shows a different picture

(World Bank 2019). In these instances, women sell cashews, groundnuts, mangoes, herbal medicine, and fish outside of the formal sector, statistics that are not calculated in terms of relative employment. In Somone, for example, I met vendors on the beach selling colorful fabric, and in Ngaparou, women of all ages commandeer fruit and vegetable stands. One of my closest friends—Marieme—ran the most popular produce stand in Ngaparou, and though she did not possess formal education beyond fifth grade it was clear that she maintained power and respect within her social and economic circles, despite operating outside of the state apparatus.

Though we will never know the real cause of such low survival rates for women onboard the *Joola* relative to the ship's overall demographics, it is impossible to ignore a discussion of the gender norms that helped shape the fate of female passengers that day (CSS 2002). What we have is a sample size of one: Mariama, the sole female survivor, was a solid swimmer who hailed from a fishing family. She was also an entrepreneur who had physical strength and a degree of economic independence. Mariema's daughter, with whom she was pregnant while onboard the ship, would ultimately survive the disaster, and upon her birth, would be aptly named "Joola" (Becker 2012).

RAISING THE WRECK: THE AFTERMATH

One of the most controversial aspects of the wreck was the government's failure to reach the ship in time to save those passengers still onboard (Technical Commission of Inquiry 2002). This is due in part to the fact that the Senegalese authorities were not aware of the crisis until the morning, for a variety of reasons that could've been mitigated with better planning (BBC November 6, 2002; Thomas and Tricot 2008). Mr. Pierre Velten, general director of the Union des Remorquers (Vessels) de Dakar, reported that he did not learn of the sinking until 0900 on September 27 when the Dakar harbor master informed him of this fact.[32] At that point, Velten said he had "prepared four tugs and called upon a team of divers," but no request for intervention was made until 1230 when he received a call from the chief of the naval staff, Commander Kombo, who declared a catastrophe and then asked him to intervene (Thomas and Tricot 2008, 33). After delays in waiting for a medical team and additional navy personnel, Velten described how the tug *Almadies* finally arrived at the wreck site at 0300 on September 28. At daybreak Velten's team discovered the bow of the *Joola*, which had floated and drifted from the original site where it capsized. He witnessed five or six containers of inflatable life rafts with only one open and inflated (High Court of Evry 2008, 33–34).

Factoring into this disaster was the fact that land based radio operators were absent from their posts overnight. Navy Commander Combo reported that Senegalese maritime officials "unfortunately did not act immediately when the ferry captain failed to make its regular two-hours calls to port" (Keaten 2002). As a consequence, those on the scene reported that the *Joola* remained capsized (but afloat) until around 1500 at which point she finally slid beneath the water's surface, taking with her those who were unable to escape the ship.[33] Though exact numbers will never be known due to unreliable manifest records outlined in earlier chapters, most agree that nearly 1,000 bodies remained trapped in the ship and were never recovered (Becker 2012). Although the International Convention for the Safety of Life at Sea (SOLAS) mandates that ships use a travel data recorder (VDR)—also known as a black box—the *Joola* was not equipped with the technology that may have helped those investigate the tragedy in detail.[34]

On October 5, just nine days after the *Joola* capsized, Smit Salvage—hired by the Dutch and led by Commander Jan ter Haar and Emil Peters—arrived in Dakar to evaluate the possibility of refloating the *Joola* wreck. The official report (Tricot and Thomas 2008, 37) recounts how the team was accompanied by an American firefighter dispatched on the basis of his experience with the terrorist attacks of 9/11. According to witnesses, only the firefighter dove the wreckage, and he surfaced after five minutes due to poor visibility, a function of swell and turbidity produced by the Gambian estuary. At this point in time, the team was able to locate the wreckage at 13 degrees 17.2 North and 17 degrees 00 East; this means that in nine days the boat had drifted more than seven miles to the northeast and entered the internal waters of Gambian territories.[35] On October 12, 2006, the Dutch report concluded that due to their inability to assess the wreck properly, Haar and Peters were unable to find breaches of maritime safety, nor recommend a proposal to raise the ship (Tricot and Thomas 2008, 37–38).

In its current state the boat now sits at a depth of 18 meters (59 feet) just 17 miles from the Gambian coast. To put this geographic location and its depth in perspective, the industry standard depth for recreational divers is 130 feet (39 meters) below sea level. During basic scuba certification, students are trained at depths of 30–60 feet (9–18 meters) and a deep dive is considered depths of more than 60 feet (18 meters). What this means is that the *Joola* wreck is so shallow that it falls within the limits of an ordinary open water dive training operation. Recreational divers could access the site to gather evidence, as could the enormous number of divers worldwide. According to the PADI organization, there are currently between 2.7 to 3.5 million active scuba divers in the United States, with as many as six million divers worldwide. I myself have the certification to dive the *Joola*, a proposal that was not supported by the Senegalese government in 2018. Despite these realities, Senegalese officials stated publicly that it was impossible to access the wreckage both then and now due to its geographical location. We know that to be false.

Immediately after the tragedy, maritime expert and diver Haidar El Ali—who has been discussed at length in this book—stated that given the shallow location of the wreckage, it was possible to drag and then raise the ferry from the ocean floor so that families could claim the remains of loved ones. As noted in the "Geography of Memory" chapter to follow, most passengers died trapped onboard the *Joola*, which in turn left many families without a necessary sense of closure. El Ali and others involved with the rescue mission waited days for the government to provide them with disaster instructions that never came. Yet the government did not waiver, even in 2005, when it had a critical opportunity to raise the wreckage from the sea (Becker 2012). According to Nassardine Aidara, head of the committee for the creation of a memorial for the *Joola* victims: "The European Union granted Senegal the necessary funds to fulfill the operation, but the authorities never allowed it" (Nassardine 2003, 44). The European Union's Eighth European Development Fund offered assistance, not just to aid Senegalese families but for worldwide victims affected by the disaster (Haidar 2018). Still, Senegal was unwilling rather than unable to meet these demands. That the ship is filled with numerous children is especially painful for the people of Senegal given the disproportionate way in which youth were affected as well as the important role that children play in West African society.

GEOGRAPHIES OF SURVIVAL AND GRIEF: VOICES FROM THE WRECK

The Diatta Family

On September 24, just two days before Senegal's 2018 commemoration of the *Joola*, I met Elie Diatta Jean Bernard, a fifty-year-old professor of finance at the University of Ziguinchor.[36] Issa Mane, my translator, managed to connect us that day to discuss the wreck and the aftermath. We visited Professor Bernard on a rainy day in Ziguinchor, made possible by the seasonal belts that bring precipitation to this tropical, verdant part of Senegal. Issa and I rolled up our pant legs, removed our shoes, and walked across tiny stones to reach Elie's compound, where he and his sister were waiting to meet us (Figure 4.3). It was a busy day leading up to the *Joola*'s commemoration, since Elie would play a central role in the day's events in his capacity as president of the ANFV survivor network. I was honored that he arranged impromptu time to meet for an interview despite the hectic day that lay before him. When I arrived, Elie was fielding phone calls, and once finished he sat and recounted the story of the shipwreck and the personal impact it had upon his family. His would become one of the most notable narratives of the *Joola* disaster, and his friendship would become one of my most important over time.

Born in 1962, his brother Michel Diatta boarded the ship that day with the task of chaperoning twenty-six schoolchildren to play a soccer match in Dakar. Michel had a lot of experience with the sport having played for several years with the Senegalese national team. Elie described how: "My brother knew everything about *futbol*. He was traveling with the children who ranged in age from ten to fifteen. In fact, the children and their families had all come to the house right before they took the ferry that day. They were wearing the same shirts, and they were so excited about the trip. Our compound was a meeting ground for events like this." Elie mentioned how his brother Michel had originally wanted to take the road for the game in Fatick, a city north of The Gambia roughly halfway between Dakar and Kaolack, but explained how his brother chose the ship instead in order to save on transport time. He also talked at length about the way in which children were affected by the tragedy: "Nearly 450 students died onboard the ship, but the largest single group of young people was actually Michel's team. Michel was so brave. He had managed to free himself from the ship but when he found out his students were trapped inside, he had returned to the boat and went down with the ferry."[37] I did not ask how the story of Michel's near survival but ultimate

Figure 4.3 Diatta Family. Source: Author.

demise was relayed to the Diatta family; both Elie and his sister seemed confident as to the truth of this narrative and it had become the local legend. Neither did it matter. As we spoke the Diatta siblings together described the intense level of fitness Michel maintained, having spent most of his adulthood as both an athlete and a coach. On the wall of their home was Michel's framed *futbol* photo, revealing both his youth and his strength. Diatta is now buried in Kabadio Niafrang in grave number 24, which his family visits every day on the anniversary of Bateau *Joola*.

André and His Family

In 2018 I also met André,[38] a first-class passenger from the neighborhood of Tildene who was one of the few souls to have survived the *Joola* wreck. At the time of the disaster he was only eighteen years old, which made him one of the youngest survivors to have endured that night, the youngest being a fifteen-year-old boy. André spoke with me in his home surrounded by his large extended family, including his mother, siblings, and numerous nephews and nieces. During interviews like this, it was typical for entire families to sit with us and listen to a narrative even though it had been told countless times (Figure 4.4). The interviews were always shared experiences, with neighbors popping in to add pieces or commentary to the story. While I was mindful of the re-wounding effect that *Joola*-related questions might have on families, Senegalese people were always open and ready to discuss the wreck and its consequences. André's family was Christian, evidence of the mixed religious traditions of Casamance, and the crucifix adorned their family compound. As we sat in a circle this young man described his ordeal though I did not press for details that would force him to revisit the tragic experience. "I swam for seven hours that night, I had managed to get ahold of a life jacket, but the water was cold, especially to be in it for that long period of time. A few of us in the water at one point saw a Chinese boat, a large one, and the fishermen asked us what we were doing in the water. We told them what happened with our ship, but they said they could not rescue us because they were not supposed to be in these waters."

Chinese vessels, as noted earlier in chapter 3, are part of the IUU network of ships that ply the coast illegally for fish, and according to André, a rescue would not be prudent and would lead to arrest. Now thirty-two years of age, André explained how it was not the *pirogues* that intercepted him and his fellow passengers, but the Spanish and Gambian vessels who were working the inshore environment that night. "They pulled us onto the ships. I was one of the youngest, but I didn't know that at the time. I wasn't a very good swimmer but I hung on." André also spoke at length about the survivor's ID, a government issued card that was distributed to those who lived through the

Joola wreck. It was not clear what purpose these ID cards served, but André mentioned his theory: "I think we have these because some people lie and say they were on the *Joola* and were not. So the documentation gives it a sense of legitimacy, to prove that we were really there."

After André recounted his story his mom relayed how she had a premonition that something was going to happen, a fact told to me by more than a dozen of those I interviewed. People described their "heart feeling heavy," or "having a dream that something bad had happened to their son or daughter," immediately before or during the disaster that night. While this notion might not fit with parts of the American worldview, it squares neatly within the context of Senegalese ideology, in which faith and tradition play a prominent role in making meaning out of events and everyday life.

Charles Gomis

In 2018, Charles Gomis told me that "if an American died on the *Joola* the whole world would know about this tragedy."[39] Charlie was one of the first people I encountered who had lost family members onboard the *Joola*. Charles was born and raised in Ziguinchor's Christian neighborhood of Tilene, one of the densely populated urban sectors set just south of the Casamance River. Charles, like many people in Ziguinchor, had purchased a ticket for the September 26 ferry, but had chosen not to travel that day for

Figure 4.4 André's Family. Source: Author.

reasons he could not explain. "I just decided not to board the ship that day, and I really don't know why." He exhibited a case of survivor's guilt or survivor's syndrome —a psychological condition that occurs when a person believes they have done something wrong by surviving a traumatic event when others did not, often feeling self-guilt. His sadness was palpable as he spoke of the loss of one of his sisters, Agnes Gomis (b. 1979), Agnes' husband Jean Pierre, and their six-year-old daughter Licalase. "My brother was living in Italy and came to visit all the family but especially my sister who was a Christian teacher and nun who worked at church. They died that day—the whole family—but the only body they found was my brother-in-law Jean Pierre, who was French and eventually buried in France."

Charles and I spoke in the back of his family compound, and in a standard show of *teranga* he insisted on giving me the only seat available in the yard, while he and Issa sat on whatever other backyard items could be found. Some members of our group simply stood listening to our talk. Charlie made it clear that for many people like him it was still difficult to process the tragedy even sixteen years after the event took place, but that it was nonetheless important to talk about it. When I asked Charlie what American students should know about the shipwreck, he replied: "That there is no justice. That the government has never helped us in the way they should. What they gave us amounts to ($22,000 USD) in concessions, and that is not enough for a life. Some of us did not take the money, but some of us had to because we had no other choice. We had a family to take care of, especially in the absence of those we lost." Issa added: "His goal would be that the government would build houses rather than give them 10 million CFA,[40] because even the money that helped the families has brought a lot of problems. They fight over the money. That's for her, not for him." While this book is not discounting the role of financial reparations for those affected by the *Joola*, Issa said that it is clear that the disaster payments created a sense of the "haves and have-nots. Most people agreed that it would've been better to help develop Casamance and support the people economically but also socially given what they had just endured."

Charles exhibited signs of depression some sixteen years after the shipwreck. Groups like the American Psychiatric Association outline how anxiety and depression following a traumatic event are natural, and that a series of steps should be taken to avoid the possible stressors of the tragedy (Jullien 2012). In the United States, victims and their families are told to "seek the advice of a psychiatrist in your community," but in Senegal, such resources are not available in villages that were most affected by the disaster. When news of the tragedy broke, more than 150 Senegalese Red Cross first aid providers were mobilized to the region to assist with rescue operations. Red Cross team leader Aziz Diop reported how "many of the families were hysterical, and it was difficult to know how to help them."[41] After all, the *Joola* event had been unprecedented, and for Red Cross workers themselves, this had been their first exposure to a major disaster.

While everyone in Casamance knew someone who died aboard the *Joola*, some families endured particularly large losses that day. This was the case of renowned footballer and Senegalese national coach, Aliou Cissee who lost twelve of his aunts, uncles, and cousins in the wreck.[42] Or, as already noted, there was the case of Coach Michel Diatta, whose twenty-six futbol players lost onboard the ship were akin to family in a nation where everyone is a "brother" or "sister" whether they are blood relations or not. The International Red Cross—together with soldiers, firemen, and other volunteer rescue teams—were tasked with comforting the families who had gathered at the port in search of missing people (IFRC 2003). Soon after, they would help the government and Senegalese families to recover, identify and bury the dead.[43] To be fair, short-term psychological assistance was in place immediately after the tragedy. But Charles and other survivors revealed the fact that many Senegalese were still processing the disaster years later, and that no formal support systems existed to help with long-term issues such as depression and anxiety. As is true in many West African families, people either endure these struggles privately, or they turn to extended families for support. The importance of social capital in regions that are economically marginalized cannot be overstated. In 2003, one year after the shipwreck, the Senegalese government put the Ministry of Health to the task of organizing a workshop to train Red Cross workers on stress management, counseling, and mourning, while examining the first responders' own reactions to "living through the tragedy." While this is a positive first step, there is still no long-term local support in Ziguinchor or rural villages for those who needed to navigate the psychosomatic issues of either surviving the event, or losing one or more of their loved ones (IFRC 2003). In 2018, I sat next to a man named Ndeye, a nineteen-year-old accountant and University of Dakar graduate who worked for the Red Cross. He was just a child when the *Joola* sank, but as we sat on the plane to Ziguinchor he described how disaster and recovery resources were "in their infancy" at the time the ship went down.

Dr. Augustin

In Ziguinchor I also spoke with Dr. Augustin and his extended family who reside in a neighborhood of Ziguinchor in the heart of Casamance. The conversation took place just a few days in advance of the *Joola*'s 2018 commemoration. While seated in their home I learned how Dr. Augustin, a lifelong physician, had been in his hospital office and his wife, Emma, was at the downtown Ziguinchor market when they first heard the news about the wreck. Both Augustin's sister Amelia, who was now fifty years old, and Augustin's son, Eric Lucien, had purchased tickets for the boat so they could get back to Dakar. Like many other young people onboard the ship, twenty-four-year-old Lucien needed to return to Cheikh Anta Diop University (University of Dakar) in time for university classes. He was studying medicine, following in the footsteps of his father, and his family had funded his studies in Dakar. Augustin described how Lucien helped him a lot on the farm in Ziguinchor, and how he loved *futbol*, and how he was just a "nice, nice guy." Both Dr. Augustin and his wife Emma had seven children, including Vivienne, Lucien's older sister by two years, who sat in the room quietly while Augustin described the events of that day. Dr. Augustin pulled out the paperwork and asked if I wanted to see the certificate of death. He continued: "The authorities had managed to find my sister, Amelia. She was a mother to four children, including three girls and one boy. She is buried in the cemetery right behind here. They managed to find my sister's body, and so they returned her to the family, so she could be buried near us. Since my son's body was never found, we say that he is in Kantene as one of the numbered graves."[44]

Dr. Augustin explained how at one point after the disaster the authorities believed they had found Lucien, but then realized that "the body was wearing *gris-gris*, which is for Muslims, and we are Christian." *Gris-gris* is a talisman, a kind of amulet originating in parts of Africa and believed to safeguard the wearer from evil. It is meant to bring luck, and in some West African countries is even used as a method of birth control. Friends in Senegal often gifted us gris-gris as a form of *teranga*; it consists of a small cloth bag, usually inscribed with verses from the Qur'an and containing a ritual number of very small objects. That so many passengers onboard the *Joola* were adorned with *gris-gris* is worth noting, because it underscores the importance of faith and tradition in everyday life and its movements across space.

Dr. Augustin also shared with me how his son had actually boarded the boat illegally, which was not uncommon, given that the ship was well beyond its passenger capacity that day. Lucien, like other passengers, had forgotten his Senegalese ID card when he stepped onboard the boat, but because he knew people who worked the ferry, they let him pass. "In some ways he forced his way onto the boat that day," Augustin declared, without judgment.[45] When Lucien's father heard word about the ship disaster, he left his office at the hospital, rode his bike home to the family compound, where family and neighbors had gathered to express their grief. "I knew in my heart that my son was dead. Everyone was crying. My wife still cries on the day of the commemoration. When I was in better shape, I would go to the water to remember, but now I just go to church. My wife's greatest pain is that they did not find our son's body."[46]

Augustin mentioned how some passengers were fortunate enough to be able to call their families to alert them as to what was unfolding.[47] This was particularly true for those who were on the top deck of the *Joola* and saw it listing before it capsized in the storm. Augustin, however, did not receive that same phone call from his son or his sister, but he nevertheless found solace in his native community in Casamance. "The way you move on is through faith, but also knowing you are not alone in this. So, so many people were affected by this disaster, so you get through it together." Augustin and his wife mentioned how one of the biggest unintended consequences of shipwreck is that "When you are dead no one can pay." I asked him to elaborate, and he described how teachers' children become teachers, or the children of doctors become doctors, and without this generation, young people cannot go on to support their parents and grandparents. It was then that Vivienne added how "many people think this was an act of God, but that is not true. This is the peoples' responsibility, not God's. There were one thousand extra passengers plus all of those cars on the ship that night." In Casamance I would hear this often, the notion of culpability in terms of the shipwreck. Many Senegalese appealed to God and faith in order to cope, but they consistently blamed the government and were in search of justice nearly two decades after the wreck.

Aliou of Ngaparou

Some people in Senegal took a more philosophical view toward the shipwreck disaster. In the north I met Aliou, a shopkeeper in Ngaparou who managed a supermarket that afforded him ample time to write his novel called *Jungle Life*. Aliou had not lost any direct relatives on the *Joola*, but he likened the disaster to so many of the other hardships his fellow citizens had faced in West Africa over time. He took a more existential approach toward the event, explaining to me that while the *Joola* was a tragedy he was more concerned with the way in which Senegalese society had not found ways to reconcile. In the preface to his handwritten novel, he writes: "We must change our mind, brain wash it, open our heart to all mankind in order to have harmony in the world. We must reject bad prejudices toward our neighbors, to be in common and lead together in the right way. For the sake of argument, we have to unite our strength, shoulder to shoulder, hand to hand in order to leave that jungle life."[48] As mentioned in other parts of this book, I encountered several Senegalese who used their spare time to craft nonfiction books or novels, in some cases lacking formal education. Aliou described how "the *Joola* was part of a larger crisis of people not coming together. The government had failed to recognize the needs of the people in Casamance, and that it was time to stand shoulder to shoulder as one country." While most people framed the issue in terms of a problem of justice, Aliou argued for the case of reconciliation (Figure 4.5). That Aliou lived far from Casamance, the region most affected by the disaster, may play a role in his perceptions of the problem and its aftermath, but it is fair to say that all Senegalese were in some way impacted by the disaster. Just as those living in Manhattan during September 11 were most impacted by the terrorist attacks, Americans in the deep south were also altered in irreversible ways by this historic event. In fact, during interviews several Senegalese likened the *Joola* disaster to the events of 9/11. The shipwreck crisis happened just one year after the terrorist attacks on New York City, the Pentagon, and DC, and therefore 9/11 was fresh on Senegal's mind when the *Joola* sank.

The Senegalese Government Inquiry into the Shipwreck

In the aftermath of the wreck, a Senegalese government-appointed inquiry into the capsizing of the *Joola* was set in motion. On November 4, 2002, the Commission D'Enquete Technique Sur issued their final report on the *Causes of the Joola Shipwreck* (Les Causes Du Naufrage du Joola), a 107-page document outlining the event and the aftermath (Documentaire 2015). The report concluded that the Senegalese army had fatally delayed the rescue operation. Three key pieces of information were released. First, Army Chief of Staff

Figure 4.5 Pirogue Responders as a Form of Community. Source: Author.

Babacar Gueye had told the government panel that there was no point in sending out military aircraft to search for survivors at 0845, nine hours after the boat was overturned in the storm. Aircraft were not dispatched until 1145 and at that time no one was rescued. Second, authorities were not alerted until the morning because radio operators were absent from their posts overnight. Third, the inquiry confirmed that given the large number of passengers overloading the ship, the ferry did not conform to international regulations concerning safety equipment (Alchetron 2018). Moreover, the panel concluded that there was a lack of training for the crew, inadequate emergency rafts, faulty radio equipment, and the captain failed to carry out basic stability calculations before leaving port (Becker 2012; Commission of Inquiry into the Joola 2002).

The Sinking of the Senegalese Ferry Expertise Report (Thomas and Tricot 2008) confirms inconsistencies in the administration of the *Joola* while laying the foundation for new changes in maritime safety moving forward. For example, the report stated that the "ministry in charge of the merchant navy was always ringing the alarm bell about overcrowding and lack of safety on board the *Joola*, without ever carrying out its legal obligation to prevent the ship from sailing" (Gounongbe 2003, 47). And, the chairman of the commission, Seydou Madani Sy, recommended that the government create a special meteorological center to provide more warnings in advance of the storm—again, emphasizing the role of weather in facilitating the tragedy. In response to this inquiry, President Abdoulaye Wade sacked Prime Minister Mame Madior Boye, who was at first blamed for dismissing any responsibility for the tragedy (VOA News 2009).

Despite this comprehensive report by the state and its admission of culpability, many remain dissatisfied with the state's handling of the *Joola* crisis and its aftermath. Interviewees repeatedly spoke of how not one person was arrested for the disaster, and there are extant demands for justice, reconciliation, and reparations by both survivors and their families in Senegal, France, and beyond. The French government has issued ongoing probes into the disaster and the Survivors network in Casamance continues to seek justice. Many interviewees saw this book project as part of the fight for justice and recognition. While the *Joola* remains in shallow waters of the coastal Atlantic, most will not be satisfied until the ferry and the full truth about the event have been raised and exposed. Only then can victims and their families move forward some two decades after Africa's greatest maritime disaster.

NOTES

1. "Lu metti yàggul te ku muñ muuñ," a Wolof proverb taken to mean "It is always darkest before the dawn."

2. Anonymous interview via Whatsapp, Colorado, USA, September 2018.

3. Anonymous undergraduate student, University of Northern Colorado, March 2017.

4. The depiction of Malian King Mansa Musa in the *Catalan Atlas*, ca. 1375, reveals him holding a golden scepter and trading gold with Berbers, illustrating European understanding of the element's origin as well as the king's power. Numerous resources on African and European images of one another may be found in Boston University's African Outreach Archives (http://www.bu.edu/africa/outreach/africa_euro_images/. See also David Northrup "Africa's Discovery of Europe" for an African-centered perspective on this historic relationship (Oxford University Press 2013).

5. Interview with survivor Ousmane Keita, Ziguinchor, Senegal, August 2018.

6. Anonymous interview, Ziguinchor, Senegal, October 2018.

7. Accurate statistics for a ship's manifest are very difficult to locate in many parts of Africa, and in some cases, do not exist.

8. It is estimated that 1,000 slave shipwrecks dot the seabed between Africa and the Americas though very few have been located (https://mg.co.za/article/2017-08-18-00-meet-the-slave-shipwreck-detective/

9. The term ro-ro is in contrast to lift-on, lift off (LOLO) vessels which require a crane to load and unload cargo.

10. Yet the *Sewol* which killed 304 passengers, including 250 high school students, has been highlighted in several Korean documentaries, including Seung-Jun Yi's "In the Absence," which was nominated for an Academy Award in 2020.

11. The maximum draft constraint of 3.10 meters was not to be exceeded upon entering the Casamance River. Draft refers to the vertical distance between the ship's waterline and the bottom of the hull (keel). This forced the naval architect to design a specific vessel with nearly unique characteristics. In 2002 only fifteen ferries had similar characteristics (in Scotland, Norway, Japan, and Greece) and among them four were night ships with cabin accommodation (Thomas and Tricot 2008, 56).

12. Anonymous interview, male shopkeeper, Ziguinchor, October 2018.

13. This ignorance of "weight in the heights" was shared by the Senegalese chief of the naval staff who declared that the commander of the *Joola* had exceeded the number of people allowed but did not reach the maximum allowed tonnage of 500. While it is true that the tonnage on September 26 was only 385, the problem was the situation of that tonnage and the resultant center of gravity (Thomas and Tricot 2008, 110).

14. BP oil has paid $10 billion USD to a businessman in a dubious arrangement in Senegal, an arrangement which has come under fire in a recent BBC documentary released on June 3, 2019 that includes accusations against President Macky Sall's family.

15. Anonymous discussion with Senegalese man, Ziguinchor, Senegal, October 2018.

16. Interview with Ousmane Keita, Ziguinchor, October 2018.

17. Interview with Ousmane Keita, Ziguinchor, October 2018.

18. The *Joola* was not equipped with a GMDSS station (Global System of Distress and Safety at Sea), was not able to receive maritime safety information disseminated via Inmarsat and could not benefit from these standard forecasts (Thomas and Tricot 2008, 113).

19. Meteo France is the French national meteorological service.

20. Mr. Maiga was interviewed for the Technical Commission Inquiry in 2002, as part of DI398, exhibit 67 of the territorial gendarme hearings, cited in Thomas and Tricot (2008, 113).

21. The Technical Commission of Inquiry into the Causes and Sinking of the Joola: Investigation Report (2002, 32–33) confirms that while rescue equipment was present onboard, the ship lacked transponders, a waterproof VHF, and a radio equipped with a DSC (digital selective calling) system which requires a single button touch to alert authorities as to a vessel's position.

22. Tricot and Thomas (2008, 66) reports that the number of lifejackets could not be determined, and it is estimated to be between 667 and 767, including fifty-eight vests for children. This was not enough for all passengers but could have saved many had they been accessible.

23. Auvray's words (2012) were translated from the French in his memoir. He lost his partner on the *Joola*.

24. Anonymous discussion with friend in Ziguinchor, September 2018.

25. Mr. El Ali Haidar determined the time of the sinking based on the fact that the victims' watches had stopped at 10:55pm.

26. Senegalese television footage showed the ferry swaying heavily as it left the southern town of Ziguinchor toward Dakar (Associated Press News 2002). This footage has not been made available to the public.

27. Wiley reported that a security guard, Omar Sow, living in the beach at Gunjar found several bodies had washed ashore during his nightly patrol twenty-four hours after the *Joola* sank, on September 27. Sow claimed that he found a man and a deceased woman, the latter with a baby on its back, both gasping for breath. He went to get help from authorities but reported that upon his return, the survivors had passed. Those survivors had to have swam 20 miles to reach Gunjar Beach from the site of the wreck, without food or nourishment (Wiley 2012, 18).

28. The author gave a talk on the *Joola* at the Walter Rodney Seminar series, African Studies Center, March 18, 2019.

29. The Navigation Permit issued on June 19, 1998, mentions 767 life jackets onboard the ferry, while the High Court of Evry lists an even smaller number at 667. Both quantities are insufficient for the number of passengers since the *Joola* was typically overloaded (Thomas and Tricot 2008, 28).

30. Anonymous interview with female student, Ziguinchor, September 2018.

31. Anonymous interview, Ngaparou, Senegal, September 2018.

32. Mr. Velten was therefore not aware of what happened nearly twelve hours after the ship sank (Thomas and Tricot 2008).

33. During the official hearing, Velten declared that the boat "ran aground" or sank at 0400 on September 30, 2002 whereas the initial report states that this happened on 1400 on that same day. These massive inconsistencies in basic chronological data

surrounding the wreck were typical and evidence of the disorganization surrounding the *Joola* rescue.

34. The SOLAS amendments, adopted on December 6, 2000, obligated ships to be equipped with safety recorders. For passenger ferry boats already in use, this modification was to be made no later than July 1, 2002, prior to the date of the Senegalese ferry disaster.

35. At that time the wreck was oriented 110 degrees by 16.5 meters deep with a pronounced list to port.

36. Author had a series of interviews with Professor Elie Bernard Diatta starting in September 2018.

37. Interview with Bernard Diatta, Ziguinchor, October 2018.

38. Anonymous interview, male resident of Ziguinchor, September 2018.

39. Anonymous interview, male resident of Ziguinchor, September 2018.

40. An article details the ways in which disputes over compensation were adjusted and delayed (New Humanitarian, July 4, 2003). See also, "Compensation of the Joola's Shipwreck Victims," in *Atlas Magazine, Insurance News Around the World*, September 5, 2013. https://www.atlas-mag.net/en/article/compensation-of-the-joolas-shipwreck-victims

41. Anonymous interview, Air Senegal flight, September 2018.

42. Aliou Cissou's soccer club in Birmingham City, England, displayed a large Senegalese flag to remember the midfielder's family, and the other people who lost their lives.

43. The government set up "crisis centers" in Dakar for relatives to help identify the victims, but many bodies were too waterlogged to be recognizable.

44. Anonymous interview with physician in Ziguinchor, October 2018.

45. Anonymous interview with physician in Ziguinchor, October 2018.

46. Ibid.

47. The average cell tower range is from 22 to 45 miles, which supports the estimate of how far offshore the *Joola* was on September 26.

48. Unpublished diary, Ngaparou resident, October 2018.

Chapter 5

Geographies of Remembrance

Memorials, Museums, and Memory

Not everything can be seen, but everything exists.

Wolof proverb.

For communities that have experienced tragedies like the *Joola*, it is difficult to build memorials that adequately depict the scale of the loss for victims and their families (Bauman 1992; Foote 2003; Sodaro 2018). Artistic approaches to rebuilding memory are challenging, because they beg large questions such as: How can the feeling of loss be properly evoked through a cemetery or museum? Who is the intended audience, the survivors, or outsiders seeking to increase their own awareness? And, in cases where a horrific event took place, should the site be reconstructed or left in ruins in order to "never forget" the incident? For maritime disasters, these challenges are particularly hard in that the physical site of the crisis cannot be demarcated nor set aside as a sacred space without logistical challenges. This is the case for many shipwrecks including the *Joola*. Yet communities can find creative ways to reconcile disasters by creating remembrances that are unique and culturally specific (Sodaro 2018). The establishment of such spaces is often bounded by geographic limitations, national and local politics, religious beliefs, and the economic resources available in that community at that time.

This chapter outlines the "Geographies of Remembrance" in Senegal in relation to the *Joola* humanitarian disaster. It covers the ways in which the tragedy has been memorialized in the form of cemeteries, museum edifices, micro-spaces of home or neighborhoods, as well as how those sites are maintained and by whom. It explores the politicized iconography of the replacement ferry for the *Joola—the Aline Sitoe Diatt*a—and how the act of naming the new boat worked to include those in the rural, disenfranchised south. The chapter also discusses how and whether *Joola* victims were identified and remembered post-disaster in line with traditional spiritual beliefs and Senegal's culture of religious coexistence. Finally, the chapter explores Casamance's frustrations with the state apparatus for mishandling of the event, while shedding light on the power of local actors to memorialize victims both during and after the tragedy.

SACRED SPACES OF SENEGAL

As a nation, Senegal boasts a variety of key memorials and sacred spaces that represent both the triumphs and tragedies of West Africa's expansive history. It is important to provide this historical context in order to better understand how *Joola* remembrances fit into the larger scheme of commemoration across the region. While it is true that memorials take time to conceive and construct—oftentimes established hundreds of years after an event—the absence of memorials also suggests something important about the politics and priorities of the time. This section begins by highlighting a few key examples of general memorial spaces across Senegal before turning to the ways in which remembrances of the *Joola* have been constructed, ignored, maintained, and by whom.

The most famous holy site in Senegal—the Grand Mosque of Touba—was founded in 1887 by Cheik Ahmadou Bamba. It is credited as the most visited pilgrimage site in West Africa, housing the tomb of Cheik Bamba (1853–1927), founder of the Mouride brotherhood and leader of the pacifist struggle against the French colonial empire. The Grand Mosque serves not only as the burial site for Bamba but as the destination for the Grand Magal of Touba, a holy pilgrimage locale for Mouride Muslims (Ngom 2016). With five minarets, three large domes, and a capacity of 7,000 people, Touba is frequented not only by worshippers but by tourists from all over the world (Babou 2005). While the mosque serves as a physical space of worship for Mourides, it is also a place that represents the spiritual work and ideology of one of the Sufi's iconic leaders.

The *Maison Des Esclaves* (slave house) is the second of three critical historical monuments, located just offshore Senegal on the island of Gorée and built to memorialize those Africans lost in the slave trade. Beginning in 1536, Gorée operated as a hub for human trafficking until 1848, when French colonial slavery was abolished. The Gorée museum was constructed within the estate of one of the former slave houses on site, and today visitors can track the history and view the "door of no return" from which slaves took the journey to the Americas. Gorée's Freedom Monument—adjacent to the museum—depicts slaves breaking free from the chains of slavery, providing a visceral experience for those wishing to understand the historical legacy of forced labor in West Africa. Built in 1962, over one hundred years after the cessation of the slave trade in Senegal, many visitors see it as an important place to remember the human toll of African slavery. Though many scholars dispute the true scale of operation at Gorée—suggesting that numbers have been inflated—it still remains an important *symbolic* site for Senegalese and international visitors, a role formalized by its designation as a UNESCO World Heritage site in 1978 (UNESCO 2019).

The African Renaissance Memorial represents a third key site of remembrance, and it is located among the central Dakar cityscape not far from Senegal's coastal zone (Figure 5.1). Just a few kilometers from the Place du Souvenir stands the 49-meter bronze statue, built for $27 million in 2010 under President Abdoulaye Wade and constructed by Mansudae Overseas Project, a North Korean company (Walker 2010). The tallest statue in Africa, it was unveiled in Dakar in front of nineteen heads of state on Senegal's National Day and designed as a prestige project that would underscore Africa's arrival into the twenty-first century on the fiftieth anniversary of Senegalese independence (National Public Radio 2010). Yet the ambitious monument has been criticized for hoovering funds that could have been better used for infrastructure, education, and health care in an already struggling Senegal economy (Barton 2017). The monument also includes a scantily clad, bare-breasted woman holding a child, which has been interpreted as being culturally insensitive to Senegal's conservative Muslim majority. To complicate matters, the president of Senegal has claimed 35 percent of the loosely defined tourism revenue, which has been met with mass civilian protest (Walker 2010).

Together these three sacred or memorial sites pay homage to some of the most important events and ideologies in West African history—Islam (Touba), independence (Renaissance Memorial), and slavery (Gorée). And while these spaces may endure ongoing criticism due to their omissions or misrepresentations, they nonetheless represent an attempt to materially manifest key events or ideologies both past and present. Given the role that the *Joola* has played in the imagination of Senegalese people, a memorial which tells the story of the disaster from a local perspective is critically needed. In this vein, thousands of Senegalese have protested the Renaissance statue's construction, and members of the ANFV-*Joola* coalition have questioned why some of those monies were not channeled into helping victims or memorializing the tragedy considering the overlapping time frame of Renaissance edifice and the shipwreck disaster.

Creating Spaces of Memory for the *Joola* Victims

The need for memorial spaces began the moment the *Joola* disaster occurred, when it was determined that government rescuers would not arrive in a

Figure 5.1 Renaissance Memorial. Source: Author.

timely manner and the death toll would mount. In the aftermath of the wreck, Haidar El Ali, who was discussed in chapter 4, brought his sixteen-member dive team to the *Joola*'s site—a journey that took seventeen hours by boat from Dakar. He recounted what happened when he reached the *Joola*'s site: "When I dove in, I saw bodies everywhere. People were screaming or hitting on the windows. We saw bodies floating by the hundreds."[1] He and his team managed to free approximately 300 corpses that were trapped inside the hull, and another 100 around the ship were retrieved.[2] All told 551 dead bodies were recovered and of that number some ninety-three were identifiable and returned to their families. Other reports added that a team of 150 military, fishermen, and rescue divers from Senegal, The Gambia, and France assisted, retrieving 360 bodies inside the ferry before "decomposition made it impossible to continue."[3] But it was Haidar, the experienced diver, who led the rescue efforts on behalf of his Oceanium team instead of the Senegalese government. Again, while the official count on the boat's manifest listed 1,034 passengers and crew, the *Joola* had nearly 2,000 passengers onboard. Many had hung onto the exposed hull immediately after it capsized, until the first *pirogues* began to encounter the vessel at around 0300 (Technical Commission of Inquiry 2002, 45; Keaten 2002). By then, those attempting to survive had been in the water for three hours given that the ship had wrecked just after midnight.

Recovered bodies were taken to various coastal sites around Senegal, many of which would be buried in informal graves along the coast or in forests. French author Patrice Auvrey—one of the only sixty-four survivors—published *Souviens-toi du Joola*, in which he outlined the poor treatment of those who died in the wreck (Auvray 2015). He described how passengers, including his wife, were trapped in the vessel when it sank off The Gambia, yet to date the ship has not been raised from these shallow, offshore waters. For those bodies that were located, they were often buried in numbered graves across Senegal and The Gambia. It was difficult enough for survivors' families to endure the loss of loved ones, but it was an even greater blow for those who were unable to find closure, having never physically identified their lost kin. Auvray underscored these thoughts in his book as follows:

> I was to be called Number 35. I was to be another anonymous victim, like all these people whose families don't know where or how to grieve. I was to be buried in the African jungle and forgotten (Auvray 2012, 23).

The recovered passengers were then placed in three main memorial sites in Senegal: Mbao, Kantene, and Kabadiou, which we turn to now.[4] While cemeteries in The Gambia also exist, they are not covered in detail here.[5]

Joola Memorials in Senegal

Given the disproportionate impact the *Joola* had on the demographics of Casamance, developing memorial spaces for the victims' families would inevitably become a subject of concern. Yet as recently as August 2018 the only existing remembrance to the *Joola* in the region was a monument in a small riverine park set in close proximity to the port of Ziguinchor. Constructed in 2003, "Le Place du Naufrages du Bateau Le *Joola*" contained a tiled fountain, anchor, and a memorial plaque to honor the victims of the wreck, though over time it has been poorly maintained and has fallen into disrepair. Locals say it is unclear who is tasked with maintaining the site, and at the time of the author's second visit to the region, the Ziguinchor memorial was set to be demolished to make way for a new, larger edifice.

Cimetiere de Kantene (Ziguinchor)

Just a few kilometers outside Ziguinchor's city limits lies the Cimetiere de Kantene, set just off the paved road stretching southward to the border with Guinea Bissau, once a destination for those fleeing rebel activity in the Casamance in the 1990s (Figure 5.2). The road passes through a verdant corridor dotted with groundnut farms, towering palm trees, and rice paddy fields, providing respite from the hectic bustle of the 300,000 plus residents in nearby Ziguinchor. Kantene Cimitiere is ringed by a cement wall enclosure and contains a total of forty-four numbered graves, and its simplicity of design speaks of grace and elegance. It is well maintained, not by the municipality, but by locals themselves. For first-time visitors the space provides a powerful snapshot of the dissatisfaction that families must face given the anonymity of those buried here. I visited Kantene for the first time in 2017 on an overcast day in August, strolling up and down the rows, observing the cement grave markers free of any identification except a number set in stone. It is a sacred and quiet space, and free of debris that is commonly found throughout the city streets of Ziguinchor. Again, names do not appear because the majority of victims were never unidentified.

Kabadio

Cimetiere des Victimes du Joola Kabadio Niafrang is another memorial site located in the Casamance region just under 100 kilometers from Ziguinchor, nor far from the Senegal/Gambia border. To reach Kabadio one must cross the Casamance River from Ziguinchor and follow the winding, paved N5 road northwest to the Atlantic coast. While many conceive of the northern region as off-limit due to the MFDC conflict, others are drawn to the space for its alternative tourism, or what some French enterprises call "solidarity tourism," a venture which began in the 1960s. Kabadio features ecotourism sites such as "Campement villageois a Kabadio," a village camp that envisions itself as culturally sensitive, environmentally conscious, and "on the fringes of the traditional tourism circuit," meant to counter the mass tourist destinations such as Saly and Mbour in the north of Senegal (Ogundayo and Adekunle 2019). Today some local "solidarity" tour groups shuttle visitors to Cimitiere Kabadio in order to witness the cemetery and honor the *Joola* tragedy. While some might argue that these tours are a form of dark tourism or voyeurism, it could also be argued that international visitors are given a rare opportunity to learn about this untold maritime tragedy through direct learning in the field.

Each year on the anniversary of bateau *Joola* the ANFV network returns to Kabadio cemetery for a private ceremony to honor the victims and families of the *Joola*. The Ziguinchor to Kabadio journey is just under 100 kilometers

Figure 5.2 Kantiene Cemetery. Source: Author.

(or two hours) via Google maps, but the trip can be much longer than anticipated due to the perils of what locals call West African time.[6] On September 26, 2018, professor and head of the ANFV, Eli Jean Bernard Diatta, gave the commemorative speech at the port of Ziguinchor on the anniversary of the *Joola*, which was replete with local and national dignitaries from across the state, including the minister of culture, Abdou Latif Coulibably, and minister of the armed forces, Augustin Tine. The ceremony took place at the pier, and the event was flanked by members of the armed forces of Senegal in full uniform who awaited the arrival of dignitaries. Victims' families were largely seated in the back of the audience, and other patient observers were able to watch as the new ferry—*Aline Sitoe Diatta*—made its symbolic way to the port. Mayor of Zigunchoir Abdoulaye Balde and many of the other politicians arrived late to the ceremony, and Diatta, president of the ANFV network, delivered the opening speech to commemorate the day.

Later I asked Professor Diatta about the politics of the ceremony, and he made clear that the most important event was not the portside observance but what followed *after* the formal festivities concluded. He was unfazed by the obvious social stratification of the elite seated in the front of the audience, and that the survivors were seated in the background. Each year, he countered, the people of Casamance caravan to Kabadio, where flowers are placed on the graves, speeches are made, and the cemetery walls painted in an effort to stave off what Diatta called the "fear of forgetting"—which is ANFV's mantra to the spirit of *Joola* memory and remembrance.[7] Diatta invited me to the event, a gesture so profound that it reminds me how the people of Casamance have welcomed others as their own despite my outsider status.

In the United States, village or municipal cemeteries are tended by sextons or caretakers, often tasked with both stewarding the graveyard and the associated church building. American states often mandate that cemeteries set aside a fixed portion of plot sales for maintenance funds, creating an implicit contract between cemeteries and consumers. In Senegal, no such resource exists, and therefore the care and maintenance of graveyards often fall on the shoulders of individuals or groups with a personal, vested interest in the site. In the case of Kabadio, the ANFV network absorbs this task and they commemorate the day outside, far beyond the realm of state or formal functions. This shows an impressive level of human agency and resistance to the larger state apparatus. Here individuals see to it that cemeteries like Kabadio and Kantene are well tended; they purchase paint or provide collective labor to honor the dead.

Mbao Cemetery

In the suburbs of Dakar sits the Mbao cemetery, where 145 numbered graves from the *Joola* wreck can be found, a site designed for affected families in the northern region of the country. Senegal does not possess high-tech forensic facilities for DNA testing, so the state had to bury many victims without proper identification (Barton 2018). Here in Mbao, Muslims, Christians, men, women, and children were laid to rest side-by-side in hastily designed cemeteries on land bulldozed by the state. In the aftermath of the wreck, some families were able to identify the dead from photos posted on notice boards, and so they were forced to witness searing images of corpses with nothing more than a number attached. Other victims were located using remaining personal effects, though the vast majority would never be identified. According to Health Ministry official Mbaye Diouf, all told, 112 passengers were identified, brought in by navy ships, of which sixty-eight were turned over to families for burial. These statistics are far from accurate, particularly for the police and soldiers who were not listed on the manifest because they rode the ship for free that day as was the custom. Likewise, children under five years of age boarded without tickets, so it is unclear what the true toll was on the *Joola* wreck that day in September.

Remembrance Square

The Place du Souvenir is not a cemetery but rather the central memorial space dedicated to the *Joola* disaster, and it is located just off the main coastal road known as the Corniche Oueste, which sits between Dakar proper and the Mamelles hills. Remembrance Square is comprised of not one but a series of squares on different levels descending as they extend toward the sea. At the far end of the memorial is a sculpture formed in the shape of the African continent, and the site itself is flanked by two pavilions and a number of blue-tiled fountains (Ogundayo and Adekunle 2019). Mourning spaces have been provided for relatives and numerous ceremonies have been held here in the wake of the disaster, given its close proximity to the capital city and associated public officials. While Remembrance Square serves as the main site for formal functions, most families from Casamance have never visited this site and probably never will (Mane 2018). This provides more evidence of the geographical distinctions between north and south, and the need for spaces of memory in Casamance on par with those found in urban Dakar.

Informal Spaces of Remembrance

While there has been ongoing frustration given the state's failure to adequately memorialize and make reparations to *Joola* families, locals have recently taken the process into their own hands, either through informal caretaking of graveyards or the design of impromptu memorials. In various sites around the country, communities have even established micro-memorials that pay tribute to the passengers lost onboard the ferry. Artwork in Fatick, for example, features an image of the *Joola* with the slogan "peace and paradise for the lost." Many of these sites of remembrance can be found across Senegal and The Gambia, and together they underscore the "fear of forgetting" mantra that the ANFV has made part of their core strategic mission of awareness and activism. Memory is a form of power and part of a community's identity, and the people of Casamance have made clear that they do not wish to forget this tragedy as the government has done.

Remembrances are also found within the micro-geographical space of the home, with shrines built to honor loved ones lost on the *Joola.* In the vast majority of the homes I visited—over forty-five in total—families reserved a special wall or nook to display photos or ephemera of their family members who were on board the ship. During a visit with Major Alpha Diatta he explained in detail the story of his seventeen-year-old son, his large photo displayed prominently next to Dia's own impressive military awards and recognitions. Diatta's son had planned to study at university when he graduated from high school but had instead boarded the *Joola* on September 26 to never return. Elie Diatta and his sister also keep a framed photo of their brother Michel on the wall, the hero who went down on the ship with twenty-six of his young *futbol* players en route to Dakar. Countless examples of micro-memorials exist across Senegal both in rural and urban spaces alike.

The Bateau Joola Memorial

The most contested *Joola* memorial site now lies in the heart of downtown Ziguinchor just a few hundred meters from the port. In some ways it provides the state's answer to "the fear of forgetting" outlined by Diatta yet the people of Casamance are not fully satisfied by its construction, as this section will soon detail. Built to replace the crumbling Le Place du Naufrages du Bateau Le Joola, the new memorial and museum will soon take the form of a boat and include spaces for remembrance and reflection (Barton 2017).[8] An 1,800 square meter parcel of land was made available by the city of Ziguinchor and in September 2018 the demolition of existing portside buildings began to unfold (Figure 5.3). I was living in Ziguinchor at this time, and I had the opportunity to watch the demolition process given my proximity to the port. On September 26, 2018, following the ceremony, the minister of culture, Abdoulatif Coulibably, symbolically laid the foundation stone for the memorial in a tribute to the victims.

The actual origins of constructing the new Ziguinchor memorial date back to 2015, when the Ministere de la Culture du Senegal launched a national architecture contest. The goal was to solicit designs for a museum that would house the *Joola* narrative and serve as a site of remembrance focused squarely on the community of Ziguinchor. The jury—comprised of two representatives of the Ministry of Culture, one representative from the order of architect, two representatives of the ANFV, one representative from Ziguinchor City Hall, and one representative of the primature—deliberated on August 14, 2015, and selected Cabinet Archi Design et Associés as the proposal winner, a design-firm based in Dakar. According to the jury the *Joola* memorial's purpose would be threefold: (1) the monument would symbolize this iconic event (*cetragique événement*); (2) it would provide a space of contemplation and prayers; and (3) it should represent the fight against the "breaches of the rules and codes of good conduct" (Ministry of Culture 2015). The criteria for the jury's decision making included the design's conformity,

Figure 5.3 ***Joola*** **Memorial Construction.** Source: Author.

the site's physical and cultural integration, key architectural features, costs, and aesthetics of the work. That the museum would recognize the need to struggle against "the breach of the rules" was seen by those in Casamance as a particularly good first step in rectifying poor behavior on the part of the state.

While a new museum might seem the ideal solution for addressing some of the long-standing omissions outlined by residents, the proposed memorial is not without its challenges and internal politics. Its imminent construction reflects some of the general dissatisfaction faced by residents of Ziguinchor. During fieldwork and interviews, locals argued that their "insights into the construction of the museum were not considered," nor were groups such as the ANFV given an adequate voice on the "site's placement" in downtown Ziguinchor. Indeed, the museum's location is central and easily accessible for those disembarking from the new *Aline Sitoe Diatta* ferry—thus increasing the possibilities of high traffic from passengers and visitors. But no consideration was given to the vendors and homeowners or other existing portside businesses before the demolition process commenced. I spoke with multiple locals during my stay in Ziguinchor and the majority of interviewees felt that the museum excluded local citizens and marginalized them in a variety of spatial and economic ways. The museum's presence may help create a level of international awareness that long been lacking, yet it is largely perceived as a failure by locals before it has begun. It is necessary to unpack this degree of dissatisfaction using interview data from my stay in Casamance in order to help understand the roots of the extant discord.

In 2018, the demolition of one large structure on the site of the proposed museum in Ziguinchor had physically displaced nearly one hundred vendors, and it represents a flashpoint for citizens of the area who have sought justice for Casamance. Local musicians, artists, and advocate Hamdine Watt mentioned how portside businesses were given an opportunity to relocate further inland, but they also sacrificed a prime real estate location which had allowed them to intercept customers arriving at port and sell their wares or services. This location also afforded easy access to fishermen working on *pirogues* stationed on the Casamance River, placing them in close proximity to the very resource they were attempting to trade or sell. In 2018 locals spoke to me about this form of "Ziguinchor gentrification," and as we talked Watt discussed how he was diligently working to find new housing opportunities for those vendors who'd been displaced. It is clear that the people of Ziguinchor care deeply about their community and are frustrated by the lack of agency in defining how the *Joola* disaster is memorialized and remembered. Yet this discontent has not stopped activists from taking matters into their own hands.

Weeks before the site's demolition I sat at a wooden table in an old two-story stone house down by the wharf speaking with a group of fishermen. I travelled many places across the country only to learn that many other people were also writing books, and this was true for Mustapha, a fisherman-turned author, who sat across the table smoking a cigarette, explaining that his house would be bulldozed in a few weeks by the municipality in order to make way

Figure 5.4 Tea (Attaya) Preparations in Senegal. Source: Author.

for the new museum. As with everywhere in Senegal, I was offered *attaya*—Senegalese tea—and I listened as Mustapha, surrounded by his friends (including Hamdine Watt) spoke of his ongoing suspicions about the Bateau *Joola* Memorial (Figure 5.4).

> We know the reason the government is building this site. It is less about creating a museum for the people of Ziguinchor, or even Senegal, and more about bringing in international tourists so that they can capture that money for the state. Chances are that the people of Ziguinchor won't be able to afford entry to the museum. Even though we're told that the fee will be free for locals we just don't believe it.[9]

The *Joola* memorial is in the incipient stages of development, and it is unclear whether Mustapha's fears about being alienated from the museum will be realized. But perhaps more important is the way in which these statements reflect an ongoing distrust of the government in terms of its willingness (or lack thereof) to provide *Joola* survivors and their families with agency or a sense of efficacy. As Mustapha's friend, a full-time fishermen, Ibrahama stated:

> We don't even know how much it'll cost to build this structure, or where that money is coming from. The state is trying to do something to recognize what happened in the wreck, and I understand that, but it lacks any kind of input from us as citizens. It seems that the building is more for them, to appease their conscience, than to help us. It is not too late to provide support for the families who lost loved ones on the ship. We still need to address that dimension, insha Allah.[10]

It is an age-old story—that of requesting recognition by the state, having the particular request granted, but in the aftermath feeling that the government has fallen short in meeting the needs of the disenfranchised. Based on interviews with locals as well as a glimpse at Ziguinchor graffiti in the downtown development district, it is clear that while people are excited about the new memorial's construction, they are dissatisfied by the results and feel excluded (Figure 5.5). This sentiment may reflect the south's ongoing disenfranchisement from the government in the north. Of the fifty-two locals who were interviewed in Ziguinchor, 94 percent argued that the Ziguinchor memorial received little to no input from the people of Casamance. When I shared with interviewees that two ANFV members in fact sat onboard the jury that determined the design of memorial, many were not content with this fact. One female interviewee explained the situation to me as follows:

> I think the point is that it's not enough. A museum is important, because it's a first step, but we continue to be left out of decisions. We have made clear that

Figure 5.5 Downtown Ziguinchor: "We want our memorial here." Source: Author.

> we seek justice for what happened, and justice has not yet been found. Since the tragedy, our needs have just not been a priority for the government and this has gone on for way too long.[11]

The cumulative frustration with the state's lack of reparations was made clear, given that 43 percent of those interviewed stated that they were willing to go so far as to *protest* the memorial's construction. When asked how that protest would manifest itself, most explained that dissent can take the form of nonparticipation, or specifically, choosing not to visit the memorial even though it would inevitably become a focal point of the Ziguinchor community. For some, the sheer cost of the museum, much like the African Renaissance memorial, represented another example of how existing funds could've been better harnessed to directly help *Joola* families and their communities.

THE LANDSCAPE OF JOOLA REMEMBRANCE AND RELIGION

Senegal possesses an undeniably unique landscape of remembrance that is rooted largely in its history of cultural and religious diversity. While the largest percentage of Senegalese citizens practice Islam, both Christianity and African world religions are common in the north to the south (Villalon

2004). These oftentimes diverging faiths also have a history of coalescence, a fact that has been supported by the Pew Foundation's work on the overwhelming level of religious tolerance here in West Africa (Pew 2010). Thus while Muslims and Christians maintain their own traditions with regard to burial rites and ritual observances, the Senegalese do not typically undermine other practices that are not in line with their faith. While in general, Muslims and Christians in Senegal maintain different cemeteries given their distinct burial traditions, it is also possible to see adherents of both traditions buried in the same funerary space. Infused in both of these traditions—Islam and Christianity—is evidence of African religions, such as the *jinn* placed on gravesites to ward off evil spirits. Particularly in the south, this spiritual coalescence is part of the larger cultural landscape found at the neighborhood scale. In the neighborhood of Tilene Kadior it is not uncommon to attend mosque on a Friday and a Catholic wedding on a Saturday, and Muslims and Christians in Ziguinchor may live across the street from a traditional African healer and seek him out for medicinal cures. It is part of the cultural evolution of coexistence that infuses all aspects of society, from everyday life, to ceremonies and rituals related to life and death.

I visited Joal, the city that birthed Senegal's first president, Leopold Sedar Senghor, where an iconic seaside cemetery housing both Muslims and Christians may be found (Kama 2010). This sacred space underscores the many ways in which the intersection between culture and religion are blurred across the state, and is particularly true in Casamance where religious diversity is the highest. In Joal Fadiouth, a lovely cemetery sits atop an island composed of shells, peppered with chunky baobab trees, and thus it has become a tourist stop for many international visitors who travel to Senegal. While a mixed graveyard like this is quite rare, it reflects what many outsiders call a culture of peaceful coexistence (Onoma 2018).

THE CHALLENGE OF THE SHIPWRECK AND SPACES OF MEMORY

While this landscape of coexistence reflects a degree of flexibility and adaptation among Senegalese people, the *Joola* crisis and the way in which it was handled by the state complicated matters related to burials. Whereas Christian and Muslim peoples are historically given a choice on where to be buried—whether in mixed or separate cemeteries—the state's handling of the *Joola* disaster *eliminated* those funerary options. Unable to identify the vast majority of bodies, these numbered victims—as described earlier—were laid to rest in mostly random cemeteries across the country. As a consequence, not only were families denied the closure that comes with identifying their loved ones, but they were unable to provide proper, religious-specific funerary rites employed under more traditional circumstances.

For Senegalese Muslims, for instance, when a person dies, a sense of urgency takes over and the body must be laid to rest shortly after death—sometimes within six hours, but often the next day if it is not possible. Over time, the Muslim funeral ceremonies have adopted some of the traditional African rituals, but for the most part they reflect the rites of the Islamic religion for the deceased (Coulon 1999). These rituals can typically last seven days and are usually somewhat complex (Eliade 1987; Onoma 2018; Wiley 2012). The body is washed and then wrapped in a large percale cloth—a closely woven, plain weave fabric. The person is then taken to the mosque for prayers, until men transport it to the grave site while women remain back at home for those in mourning. Post-burial kola nuts—highly caffeinated and used at naming ceremonies, weddings, and funerals—are distributed to mourners, while visitors themselves provide the bereaved family with financial donations to assist with the cost of the proceedings (Hayden 2009). While there is some variation across Senegal given the need for flexibility, some of these traditions must maintain a certain rigidity in order to adhere to the tenets of Islam. That the *Joola* left an enormous number of bodies unidentified would inevitably interfere with these traditional burial customs, causing yet another psychological blow to families working hard to recover from the disaster (Levtzion et al. 2000).

Christian burials—though less common given that Senegal is a 95 percent majority Muslim country—adhere to certain and specific customs during funeral processions (Jindra 2005; Langewiesche 2011). Relatives are buried in caskets rather than cloth, blessed with holy water by a priest, and a church service (given in Wolof) is arranged in order to commemorate the person's life and Christ's death. Not unlike a Muslim burial, food is shared with guests, and in a show of reciprocity, those mourners donate money to assist with the family's loss (Onoma 2018; Klein 1968). During my focus interviews with fishermen in Ngaparou, a coastal city in the north, the issue of Joola burials came up repeatedly, though it was not part of the original semi-structured question set. A part-time fisherman and construction worker named Babacar elucidated the problem as follows:

> I don't know the casualty numbers for sure, because we really don't trust what the government tells us about the *Joola*. Many of us don't care about the money they gave families, because the government can't make reparations for what we lost in terms of human life. But in failing to react quickly and purposefully to save those onboard, we faced other losses. We will never know who is buried and where. We were unable to give proper burials to those we lost, and this is something that will never, ever be reconciled, in my opinion.

> I'm not afraid to speak out about this because it is the truth. I don't know how to mourn. I don't know if our families are at peace because there was no proper ending.[12]

As Babacar suggests, as with any unexpected tragedy unintended consequences will result, with long-standing repercussions for the families and friends left behind. In some ways, the state had no prior experience, no benchmark in which to address such a large-scale humanitarian disaster, so inevitably, mistakes would result as the country navigated its way through the tragedy. But despite Senegal's religious flexibility evidenced by the mixed graveyards such as Fadiouth or religiously diverse neighborhoods like Tilene Kadior, these micro-transgressions affected people deeply. Senegalese and, for that matter, international victims not only lost their lives in an avoidable tragedy but also were denied the ablutions of death in line with their faith-based traditions. In August 2018, I sat with Thierno, a twenty-year-old shop worker in Ziguinchor, discussing the *Joola*. Sitting in his father's boutique, we talked nearly every day, while he sold cigarettes and phone cards and snacks to locals, of which he knew all by name:

> I was very young at the time of the Joola. We've faced tragedy here in Ziguinchor, so we are accustomed to that. But I know that to be denied a proper send-off, or to leave life without the rituals we've seen and experienced, is very sad.[13]

The reality was that many Senegalese families adapted to these unconventional circumstances by holding symbolic rituals within the space of home. They continue to hang photos on their walls of lost children, uncles, and aunts, or even friends who boarded the ship that day. In other cases, they visit nearby cemeteries, whether relatively new ones such as Kabadio or Kantene, or existing ones, and thus pay tribute in a more generalized way. Unable to view the site at which their loved ones are at rest, they practice their rituals in the best way they know how, through their ongoing flexibility and willingness to adjust to life's changing circumstances.

In the absence of a body, the Jola, the largest ethnic group in Casamance, hold an animist ceremony that allows them to recall the spirits of the dead (Nassardine 2003; Thomas 1982). This was the case in Mlomp, where fifty-four students perished, not one of their bodies recovered from the corpses that were found and brought home. The Jola ceremony begins on the second or third day of the seven-day funeral service and is conducted in the evening. In the case of male deaths, men of the family and one male from each of the districts assemble in the village center. From there, they walk to the border of the sacred forests, facing the direction where the victim had last been seen.

They sing songs, usually ones that were favored by the victim as a child, or they tell a story through song. The songs are like a clarion call for the victim, and they sing at the edge of the forest until they believe the spirit has arrived. Once the assemblage is certain that the victim is nearby, they turn back to the village center, continuing to sing. They do not look back and, similarly, they extinguish any fires for fear that the deceased will be blinded or lose his way back to the forest. Then, the group sings by the altar, with the goal of raising the spirit of the deceased to an ancestral heaven. The same tradition is practiced for women victims, with women forming the choir and assemblage. While this Jola tradition and most others do not require a body, the "Question and Answer" ceremony does (Levtzion et al. 2000).

The question and answer ceremony is launched by the individual closest to the deceased. This person announces that the body is ready to be carried outside the house for the ceremony, usually on the second day. The body is then placed on a stretcher surrounded by four pallbearers who have not met, or in contrast to American customs, knew very little about the deceased (Diawara 2000). Once outside, the mourners form a group, facing the pallbearers, the latter of which feel an invisible force leading them to one individual (Leary 1970). This person is then selected to ask the questions of the deceased, such as "Was it a good spirit that took your life?" The pallbearers move back by way of the invisible force if the answer is "yes." If the deceased is thought to have a good soul, it is believed that he or she will move onward to an ancestral heaven. If not, the soul will be embodied in a lower animal form such as a hyena. Many *Joola* victim's families had two ceremonies—a Muslim one, as well as a traditional African one like that described here, keeping in line with their traditional origins. The sheer number of deaths in Casamance meant that families had to endure numerous week-long funerals after the *Joola* wreck. Since Senegalese see everyone as their brothers and sisters, even if they are not blood-related, the grief was overwhelming for so many in the region who had to repeatedly experience the pain of loss.

The Jolas are the most dominant ethnic group in the area, and many have rejected both Islam and Christianity. There is also not a hierarchical culture like many other ethnic groups in Africa. The Jolas are as a result fiercely independent and many have participated heavily in the MFDC rebel movement in Casamance. While animistic in general, the Jola, like many Africans, are more complicated in terms of their faith-based traditions. Rather than engaging in Sunday or Saturday rituals, for the Jola the whole of existence is a religious phenomenon. Mbiti points out how Africans believe that God is the source and ultimate controller of the force or energy that permeates the whole universe (Babou 2005; Reese 2014). The spirits have some of this energy and a few individuals have the knowledge and ability to tap it—thus we see the creation of medicine men or marabouts, some for good, and some for evil. For the Jola, any religion that tries to replace the traditional African religions must occupy

Figure 5.6 ***Aline Sitoe Diatta*.** Source: Author.

the whole person as these traditional religions do. Thus for many Senegalese, particularly the Jola in Casamance, religion is used to understand and explain the world and one's place in it, not just at weddings and funerals, but within everyday life (Klei 1968). So the *Joola* disaster, for Jola people, was not just a mechanical disaster but an event that was tied to good and evil and explained as such. The failure of the government to return bodies to bereaved families complicated the norms of these cultures, rendering them unable to pay proper respect and make meaning out of the greatest tragedy of their lifetime.

A REPLACEMENT SHIP AND ITS NAMING

In addition to the human toll of the ferry wreck, Senegalese people also had to contend with the disruption of the only maritime transport route connecting the country's north and south. Economic and cultural ties between these regions were, and are, critical, and thus there was a pressing need to resume the transport of products and people in a quick, affordable manner. The Senegalese government therefore had to replace the ship with a new, seaworthy vessel capable of carrying passengers between Casamance and Dakar. Travel between north and south was still an option by road, but fraught with high costs, time delays, and safety issues, owing to the politics of The Gambia and the MFDC situation in Casamance. While many residents had no interest in boarding a vessel after the *Joola* wreck, others felt the need to carry on with the economic activities of everyday life. Thierno's father, Ibrahima, summarized his thoughts as follows:

> Of course we were very scared of what happened and frightened to board a ship again. But most of us have family in the north, or if not, at the least, we are selling things to the people in the north. The supplies for my market come from the north, so at some point, we need to move on. We survived the *Joola* but we also need money to survive life.[14]

This need for financial support was exacerbated, in some cases, by the loss of passengers on board the *Joola* who had served as wage earners for their families. Or, in other cases, where children or youth were lost in the wreck, families had to contend with the loss of *potential* family income down the line. Family networks are undeniably strong in Casamance and dependent upon such economic and livelihood integration. The loss of a wage-earning family member can have important social repercussions but also negatively impact the financial sustainability of the entire family unit.

In 2008, the answer to this dilemma came in the form of the *Aline Sitoe Diatta*, a ship named after the Senegalese heroine who led the opposition against the French colonial empire during World War II (Wilmetta 2005). As the Jola leader of a spiritual group in Kabrouusse, Aline Sitoe Diatta (1920–1944) was often called Senegal's Joan of Arc, in part because of the compelling way in which she responded to the French government's seizure of the region's rice harvest in 1942 (Baum 2015). Diatta was part of a women's group that led a boycott against these resource incursions in the 1940s and, as a result, she was imprisoned before being deported to Timbuktu in 1943 (Toliver-Diallo 2005). Though she died at the age of twenty-four in a Malian jail due to disease on May 22, 1944, she has become a symbol of resistance in West Africa, in general, and in Casamance, in particular. In addition to the ferry, numerous sites, schools, and businesses are christened after Diatta, known as "the woman who was more than a man" (Baum 2015). Tributes are extensive across the region and even infuse popular culture, and in 2016 the Swedish-Senegalese duo "Sousou and Maher Cissoko" even produced a song in tribute to this Senegalese heroine.

> Aline Sitoe Diatta, You are the one we choose and
> you are the one we see, Aline Sitoe
> You are a woman with more power than men. You can win.[15]

That the new ferry line's name was given to a female was noteworthy. But the fact that the ship was christened after a female freedom fighter who hailed from *Casamance* was more pressing, given the disproportionate impact the shipwreck had on southern people and the government's need to make, at the very least, symbolic amends to a marginalized people (Toliver-Diallo 2003). Though Diatta's story is often employed in a way to promote Senegalese nationalism against the French, she was very much a figure and icon of the rural south (Konate 2003).

The Replacement: The *Aline Sitoe Diatta*

The new ferry—the *MV Aline Sitoe Diatta*—was first constructed by Fr. Fassmer Shipyards in Berne, Germany, between 2006–2007 and began active ferry service in March 2008 along the Atlantic coastal route between Dakar and Ziguinchor (Figure 5.6). The ferry is now owned by the Ministere d l'Economie Maritime et des Transports Maritimes Internationaux and operated by COSAMA SA (Consortium Senegalais d'Activités Maritimes 2018). Built to accommodate 504 passengers, thirty-nine crew members, and twenty-eight vehicles, the scale of the ship is not unlike the *Joola*, though the ferry has been given far greater security measures in the wake of the 2002 tragedy. Similar in dimensions to the *Joola*, the *Aline Sitoe Diatta* was designed as a ROPAX—or roll-on/roll-off passenger ferry—built for freight vehicle transport and passenger accommodations. As a 3,481 gross ton ferry (length: 76 m; beam: 15.5 m; draft: 3.2 m), its dimensions are roughly the same but with an increased cruising speed of 14.5 knots and a range of 1,750 nautical miles.

In 2017 and 2018 I spoke with locals about whether the naming of the replacement ferry helped them to make meaning out of the disaster, whether the symbolism of Diatta was a genuine concession to help unite the country, or whether it mattered at all given the circumstances. The replacement ship had been installed roughly six years after the wreck, so the state had had ample time to consider the optics of christening a ferry after a female Casamance freedom fighter. When *Aline Sitoe Diatta*'s name was mentioned with interviewees, it certainly evoked enormous respect from Senegalese people, regardless of their gender, ethnicity, or residence in the north or south. Yet as in most discussions, local people could not look past the failures of the state apparatus in handling the crisis to see the ship's naming in a positive light. It was clear that they respected the warrior Aline, but not the government, despite its efforts to make symbolic concessions. As one thirty-five-year-old female employee of Hotel Kadiandoumagne noted:

> The naming of the ship does not matter at all to me. If you remember, the *Joola* itself was named after the Jola people here in Casamance, and look what happened.[16]

The *Aline Sitoe Diatta* as an Unintentional Site of Remembrance

In light of this chapter's theme, "Geographies of Remembrance," I argue that the *Aline Sitoe Diatta* itself serves as the most important yet unintentional memorial to the *Joola* shipwreck of 2002. There is no better way to understand the fate of *Joola* passengers than to board this near identical ferry and travel the long overnight coastal journey between Dakar and Ziguinchor. The experience provides an opportunity for locals as well as outsiders to embody what remains, to this day, Senegal's greatest humanitarian disaster. Though not designed by the state for these purposes, the *Aline Sitoe Diatta* serves as a space in which to remember the fate of friends and relatives in their last hours on the ferry. And at the same time, the journey enables passengers and survivors to revisit the positive experiences they had on board the *Joola* in days past, traveling between Dakar and Casamance with friends and families.

My conversations with locals when I rode the new ferry confirmed how the *Aline Sitoe Diatta* is indeed a critical space for memory and reflection. In 2017 Amina and her two children sat portside on a wooden bench, sharing their evening meal of French fries with me as we headed south overnight toward Ziguinchor. As the children listened to their mom, Amina spoke about her childhood memories of the *Joola*:

> I used to travel on *Joola* with my family to Dakar, and we would buy third class tickets and stay up all night on the top deck. I have wonderful memories from the boat. It is so sad what happened here. All of us were affected but those days were good. We all lost family members.[17]

My colleague in Dakar recalled similar experiences on the *Joola* when he would travel regularly between the university in Saint Louis to visit his family in Ziguinchor. He shared photos of his friends huddled together on top deck, in what was clearly a joyous experience for he and his contemporaries who were out on a university break:

> It was a very fun experience. I have so many good memories of the boat, it's still hard to believe that the *Joola* is gone when I look at these photographs. But we have to move forward and that's partly why some of us travel on the new boat now. There is no other choice.[18]

Traveling this long night journey between the two major cities in the north and south, it is impossible not to imagine the fate of the ship and empathize with its passengers. While such visceral, emotional experiences might rewound those who survived the tragedy, it is clear that a *moving memorial* also has its benefits. For insiders like the Senegalese, the ferry ride enables them to process the shared, positive experiences they had with loved ones while traveling the Atlantic coast at night. It is an undeniably singular experience to travel the overnight ferry and to arrive in Casamance by mid-morning, witnessing groggy market sellers preparing to load mangos in anticipation of the boat's arrival. And for outsiders, the *Aline Sitoe Diatta* journey inserts us into the shoes of the Senegalese people on the night of the wreck, allowing us to better understand this great disaster with increased understanding. It is one thing to read about a disaster, but altogether different to witness the site of tragedy and to meet the people most affected by the event. Such memorial spaces, whether intentional or otherwise, help us to make meaning out of tragedies and better humanize the communities that were directly affected by the loss. They also fit squarely with the Sufi tenet of invoking optimism in the face of tragedy and despair.

NOTES

1. Interview, Haidar El Ali, Senegal, 2002.
2. As noted this is not a disaster narrative. However, there are a few video resources for those interested in witnessing the *Joola* and associated tributes. https://www.youtube.com/watch?v=TCnxAQRnWJg
3. The Associated Press reported on the rescue operation and its delayed response in 2002. https://apnews.com/a406d5ec772133d2d3bbf05299921f8d
4. Film footage of "Rideau Rouge sur le naufrage du Joola" may be found in French: https://www.youtube.com/watch?v=-3pMgLCvsAY&feature=youtu.be
5. According to Haidar and other accounts, nearly 1,000 victims remained trapped in the boat, never to be recovered (personal communication 2018).
6. Not only does Senegal sit within the Greenwich Mean time zone but it is also subject to *West African time*, a Senegalese colloquialism that refers to the fluid nature of appointments and schedules. A thirty-minute interview might transform into a full-day affair, but bring with it serendipitous opportunities to forge new connections with village chiefs or mango sellers.
7. Interview with Professor Diatta, Ziguinchor, October 2018.
8. A government official stated in 2018 that the memorial will cost between two and three billion CFA.
9. Anonymous interview, male writer, Ziguinchor, October 2018.
10. Anonymous interview, male fishermen, Senegal, July 2018.
11. Anonymous interview, female resident, Ziguinchor, August 2018.
12. Anonymous interview, male resident, Ziguinchor, August 2018.
13. Anonymous interview, male shopkeeper, Ziguinchor port, September 2018.
14. Anonymous interview, male shopkeeper and owner, Ziguinchor port, October 2018.

15. Sousou and Cissoki's video is found here https://www.youtube.com/watch?v=bII8XqGjLQ4.
16. Anonymous interview, female hotel employee, Ziguinchor, August 2018.
17. Anonymous interview, female stay at home mother, Aline Sitoe Diatta, August 2017.
18. Anonymous interview, former Ziguinchor resident, phone conversation, April 2019.

Conclusion

Geographies of Omission and Optimism

It is better to walk than curse the road.

Wolof proverb

When I presented research on the *Joola* disaster at academic conferences and in classrooms, the majority of audiences informed me that they had never heard of Africa's greatest shipwreck, let alone knew about the impacts it had on Senegalese society. Audiences would frequently blame the lack of awareness on the engrossing and exhaustive news cycle of 2002, which coincided with the timing of the wreck.[1] Colleagues, students, and friends would speculate that because the ferry disaster happened just one year *after* 9/11, the United States had "turned its attention elsewhere" or "inward." This implied that our focus had been on the African continent to begin with, a myth that will be dispelled later in this final chapter. The events of 9/11, in which terrorists assaulted our nation and its ideologies, created unparalleled consequences for society, and the scale of this event necessitated the full attention assigned to it by the media. On October 7, 2001, the United States responded to the 9/11 attacks with *Operation Enduring Freedom* and the *War in Afghanistan,* followed by airstrikes targeting both Al Qaeda and the Taliban.[2] That said, on September 26, 2002—the date of Senegal's shipwreck, and just one year after President Bush Jr.'s declaration of the *War on Terror*—America was grappling with a "new normal"[3] and had limited bandwidth in which to pursue other stories, particularly those that did not align with its core strategic mission.[4]

A content analysis of news media coverage in the United States confirms how the shipwreck was grossly omitted or altogether ignored in the press (Mericle 2016). In September 2002, and in the months and years to follow, there were only a handful of U.S. articles pertaining to the *Joola*, but their numbers were not proportional to the magnitude of the event. For example, on February 20, 2003, *New York Times* climate correspondent Somini Sengupta wrote, "Africa's Pride and Joy Vanished When Ferry Sank," outlining the events of 9/26 and its effect upon younger demographics across Senegal. Sengupta accurately described the resulting brain drain the wreck had on Casamance, while detailing what that loss meant for African societies, particularly since youth are viewed as a vital link between past and future (Sengupta 2003). On February 23 of that same year, the *New York Times* followed up with a second, shorter piece entitled, "Deaths at Sea: Grim Numbers," which, in 108 short words, ranked the worst non-military disasters of all time. The article accurately slotted the *Joola* ahead of the *Titanic,* yet it did not lead to deeper, more sophisticated coverage of the wreck (*NYT* 2003). Together these two articles formed the extent of U.S. east coast media stories about Senegal's shipwreck. Even on September 27, 2002, the day after the *Joola* capsized, the front page of the *Times* would include multiple stories related to 9/11, the GOP house race, Russia's rebel fight, a Congressional resolution on the force in Iraq, and a feature on the University of Mississippi riots of 1962 (*NYT* 2002; September 23, A1). There was nothing whatsoever written on the Senegalese disaster. While readers might find this hard to believe, the *Joola* shipwreck was simply not on America's radar in the years surrounding the war nor in the decades to follow. And as discussed in the Introduction, even the popular magazine, *National Geographic*—a publication best known for its global coverage—has not written about the *Joola* in the eighteen years since the wreck (Mericle 2016). As of the completion of this book, there is currently no existing record of the *Joola* shipwreck in any of the *National Geographic*'s print or interactive media. Pat Wiley (2012) would prove to be the only American until Barton (2017, 2018) to provide detailed, investigative fieldwork that went beyond disaster narratives and *Joola* sound bite.

Across the Atlantic, British journalists provided greater coverage of the *Joola*, though their content focused primarily upon the ferry's death toll and called out those officials most culpable in the disaster. Articles in the *BBC World News* briefly highlighted the role of Senegalese officials, both government and military, who were negligent in responding to distress calls from the boat (BBC World 2002, 2003). The French, in contrast, who lost twenty-two nationals in the wreck, provided the most comprehensive treatment of the event outside of Senegal and The Gambia,[5] where the disaster is still discussed in journalistic circles, given its enduring impact on its people and economy. The French produced countless media articles on the wreck as well as several excellent memoirs that paid homage to the victims and their families (Auvray 2012, 2014; Becker and Diallo 2012; Eusebio 2005; Gounongbe 2003; Nassardine 2013). Ties between Senegal and France remain strong, owing to the former imprint of colonial rule; many of these shipwreck memoirs were thus authored jointly by those of French and Senegalese heritage. Given that five Spanish nationals, mostly tourists visiting Casamance, also perished aboard the ship, there was some minor coverage of the wreck on news websites and blogs across Spain (Jar Torre 2003). In short, America's coverage of the *Joola* paled in comparison to stories elsewhere, particularly in Europe where several passengers had lost their lives. There were, as previously noted in this book, no Americans onboard the *Joola.*

THE FALLACY OF PREOCCUPATION

Yet the lack of U.S. media coverage about this tragedy cannot simply be explained by America's preoccupation with its War on Freedom in the year 2002. Neither can the media void be fully understood in terms of the absence of Americans on the Senegalese boat. After all, on November 19, 2002, the Greek-operated, Liberian-registered ship, the *MV Prestige*, sank 130 miles off the coast of Galicia in a storm, just *two months* after the *Joola* wreck occurred, and between 2002 and 2012 alone the Spanish incident was covered fifteen times in the *New York Times* (NYT archives). In 2002, the *Economist* also published "The Prestige Oil Spill: Game of Consequences," which detailed how 17.8 million gallons of heavy fuel oil[6] leaked from the structurally deficient oil tanker, contaminating coastlines in Galicia, making it Europe's worst environmental disaster. The World Wildlife Fund's senior policy officer for shipping stated: "The environmental devastation caused is at least on par, if not worse, than the *Exxon Valdez*. The amount of oil spilled is more than the *Valdez* and the toxicity is higher, because of higher temperatures" (Vince 2003). Geographers Pulsipher and Pulsipher (2008) would also see the educational value of including the *Prestige* case study in their *World Regional Geography* text for American undergraduates (Pulsipher and Pulsipher 2008, 178). Their 2008 chapter vignette includes an account of the maritime incident and a link to the PBS *Frontline* video, "The Lawless Sea," which brings the ship's story to life for college students.[7] Together, these resources help American audiences to better understand the international profiteering that gave rise to the event and its effects on marine wildlife. Yet the *Joola* shipwreck has never been afforded the same media attention assigned to the *Prestige*, despite its profound impact upon the mortality and livelihoods of Senegalese *people* relative to the Spanish ship.[8] Though the *Prestige* wreck represented Europe's worst environmental disaster, the Filipino crew were all brought to safety, evacuated in the aftermath by rescue helicopters. Not one human was harmed in the Spanish disaster. Nor was one single American onboard the *Prestige*, but Europe's story still captured our collective interest at that time.

THE *MV SEWOL* AND THE MYTH OF AMERICAN APATHY

Evidence supporting the fallacy of America's preoccupation with its *own* news stories mounts upon closer examination of the case of the MV *Sewol* disaster and the marked focus it drew (and continues to draw) from the press. As noted briefly in chapter 4, the South Korean ferry wreck occurred on April 16, 2004, killing a total of 304 passengers including *250 students* from Danwon High in Anson, not far from Seoul (Lee and Sang-Hun 2014, Section 1, Page 6). Not unlike the *Joola* ship, the *Sewol* was constructed as a ro-ro passenger ferry and boasted a gross tonnage of 6,835 and an imposing length of 145 meters.[9] It was larger in size and speed than the *Joola* but similar in purpose and overall design specifications. Like the *Joola*, of the approximately 172 *Sewol* survivors, more than half were rescued by fishermen who served as first responders; though in the case of the Korean ferry, the Coast Guard also arrived at the scene, in contrast to the notable absence of maritime authorities in Senegal. While there was no storm implicated in the Korean incident, the *Sewol* had also taken on twice the maximum allowable cargo load. The ship was further compromised by the insufficient quantity of ballast onboard (761 tons) and the failure to maintain ballast tanks or properly secure cargo (*BBC News* 2017), and so there were numerous similarities between both of these ferries given their precarious status.

Beyond ferry specifications, the most important parallel between the *Sewol* (2014) and the *Joola* (2002) was the sheer number of youth lost once the ships capsized. Some 450 youth died on the *Joola* (Barton 2017; Becker 2012; Wiley 2016) and as noted, 250 high schoolers were tragically lost when the *Sewol* sank. For our purposes, the Appendix[10] provides a closer look at the large quantity of youth who perished on the Dakar-bound ferry in 2002. While the *Sewol* predominantly affected high schoolers, the *Joola* had an astonishing impact upon all cohorts aged twenty-three years and younger (Becker 2012; Chenore 2019). A deeper dive into the *Joola* data reveals that 521 persons from the 18–29 year cohort perished; 154 did so in the 12- to 17-year-old cohort, and among passengers under 11 years of age, at least 213 died. A quick glance at the data illustrates that *around 110 of the victims were five years of age or younger.* These statistics do not account for the bodies who were not assigned ages due to identification and accessibility issues; yet locals speculate that a large proportion of these were also likely five years and under, since that cohort boarded free and were thus not included in the ferry's manifest. Poor recordkeeping onboard the *Joola*, the lack of post-disaster DNA testing, and the sheer number of bodies still trapped on the boat prevented the release of accurate statistics for age.[11] However, it is difficult to argue that a tragedy involving 223 children, ages eleven years and below, under any circumstances, should ever have been ignored by the international press. In the same way Senegalese officials were negligent, so, too, was the global media in a crime of omission and apathy.

In sharp contrast, the *New York Times* has covered the *Sewol* incident in depth, with a total of twenty-four archived articles, the bulk of which focus on the loss of Korean youth (*NYT* 2020). There have also been multiple documentaries produced in the wake of the disaster, and while most of them originate in South Korea, these films have been elevated by American media. This includes Seung-Jin Yi's *In the Absence*, which was nominated for an Academy Award in 2020. In 2019, the *New Yorker* released, *What Went Wrong in the South Korea Ferry Disaster?*, a documentary replete with footage from dash cams on the ship, and interviews with parents, survivors, and first responders. In 2018, director Kim Ji-Young produced *Intention*, which shattered box office records not just in South Korea but also abroad for its detailed coverage of the wreck. Sang Ho Lee and Hae Ryong Ahn's *Diving Bell: The Truth Shall Not Sink with Sewol*, like the other exemplary *Sewol*-themed films, underscores the stories of high schoolers impacted by the tragedy.

These are not disaster narratives, but rather comprehensive investigations that go beyond the casualty count to understand the root causes and consequences that the *Sewol* crisis has for Korean society. Lee and Ahn, in particular, use original footage to reveal how media outlets in Korea colluded with government officials to release false information that would conceal the extent of the tragedy (Asia Press International 2014). *Diving Bell* was produced just six months after the *Sewol* sank off the coast of Jindo Island; due in part to pressure from filmmakers and families, the ferry was eventually raised from its depth of 144 feet just three years after the ship went down. Given that the *Sewol* sank just six years ago, it is astounding that multiple documentaries have already been made on the subject in such a short period of time. Suffice it to say that despite its enormous impact on Casamance youth—which form the core of African society—the *Joola* has clearly never been given that same level of attention or due respect.

I am not attempting to ignite a *competition over casualties*, pitting the *Joola* and the *Sewol* disasters against one another. Both are important and critical narratives in their own right and must be disseminated to the global public. Rather, the *Sewol* example is employed here to elucidate the fact that while Senegal's shipwreck surpassed the Korean ferry in terms of its human toll, it has been virtually ignored by mainstream press. Some might hypothesize that Senegal, given its limited resources, does not have the same capacity to produce films as South Korea does, given the disparate human development indicators and GDP numbers in these states.[12] In other words, some argue there have been more resources available to tell the *Sewol* story.

But this does not provide a complete nor comparative picture of the situation. Despite economic setbacks, West African states have an undeniably rich and burgeoning film industry,[13] evidenced, for example, by the work of Senegalese director, Moussa Toure's *la Pirogue*, and his effort to shine a light on poverty, outmigration, and the geographical space of the sea. Here, filmmakers are exploring the complex stories that debunk the dominant discourse of Africa as a space of warfare and disease. African directors have the capacity to tell African stories, too, if not alone due to their resource constraints, then in collaboration with European, American, or Asian artists. Thoughtful African-centered collaborations are particularly important in light of the controversy surrounding *American Dirt* (Cummins 2020), a book which has been criticized for positioning immigrant stories through a non-immigrant lens.

THE MYTH OF EXISTING MEDIA IN SENEGAL

One other rationale used to explain the void in *Joola* storytelling is the limited availability of original photos, video, or audio from the shipwreck in contrast to other, more publicized wrecks, materials that might help foreign audiences better visualize the Senegalese tragedy. The 2014 *Sewol* documentary *In The Absence*, for example, includes audio from passengers and emergency operators right after the ship capsized; live video footage of children being told to "stay put" by ship operators; and calls with the Jindo Vessel Traffic Service and Maritime Safety Division that revealed government ineptitude. The film, *What Went Wrong in the South Korean Ferry Disaster?* also incorporates powerful footage of the ship's captain boarding Patrol Boat 123 before his own passengers even commenced evacuation (*New Yorker* 2014). *Sewol* survivors like Kim Soon-Mook are interviewed, text messages from students flash across the screen, and clips from dash cams in the ferry show the cargo hold being flooded. Much of this original footage is underlaid by more recent drone videos that return to the scene of the disaster. In sum, there is ample data to help humanize the shipwreck story for audiences who have no personal connection to South Korea. Some argue that by comparison there was insufficient data to roll into a *Joola* film or magazine article that would appeal to international audiences, yet that is not the case.

This argument crumbles upon closer analysis of *Joola* visual or audio archives relative to those of other historic wrecks like *Titanic*, the latter of which is now a global legend. These two disasters are separated by nearly a century, yet historians have managed to accumulate an impressive body of literature and media highlighting the *Titanic* despite the lack of original, existing footage from that time period. Lest we assume a false equivalency, the Senegalese and British stories possess key geographic and temporal similarities and are worth discussing in tandem. The *Joola* was roughly twenty miles from shore when it capsized, and the *Titanic* was even further from the coast of Newfoundland at 400 miles. Both disasters began to unfold around 11:00pm, and this evening time frame coupled with the distance from land complicated both of the rescue missions.[14] But in terms of the *Titanic*, Americans have internalized much of the details of that evening by way of films and books. Yet it often surprises people to learn that the only bona fide clip of the RMS *Titanic* shows her being towed into the outfitting wharf at Belfast in February 1912. In fact, footage of *Lusitania* leaving port has often been shown as a substitute for *Titanic*'s maiden voyage commencement.[15]

That we know so much about the New York-bound *Titanic* is a function of careful, painstaking, archival scholarship combined with interview data and advances in modern-day exploration. Director James Cameron, for instance, dove the *Titanic* wreck a total of twelve times to capture the ship's authenticity, footage that has been dovetailed into documentaries and museum exhibits worldwide (*The Titanic Experience* 2019). In 1912 the world lacked the technologies to disseminate the *Titanic* story in real time (texts, video), yet modern scholars managed to cobble together detailed evidence to tell this important story. We've seen so many visuals of the ship that it's as if we, too, were on board. In contrast, the *Joola* tragedy happened in 2002, when the availability of media was actually more robust, particularly in terms of mobiles and text messaging functionality. The Pew Report (2015) underscores this fact; African cell usage has been largely on par with that of the United States since the 2000s. West Africans were early adopters of these technological innovations, allowing them to leapfrog across the digital divide using inexpensive mobile phones. During my interviews in Casamance I was told of the frantic texts that parents received from their children onboard the *Joola* the night it capsized. These texts from passengers, like those from the *Sewol*, form a key part of the *Joola's* archival record. Furthermore, noted diver and ecologist El Ali Haidar captured footage of the ship immediately after it capsized, providing clips that could be folded into a compelling story.[16] In sum, for the *Joola*, there is certainly enough visual evidence to recreate the shipwreck story for audiences, but we need to begin creating and then disseminating the work.

THE RMS *TITANIC* AND *LE JOOLA:* OPPORTUNITIES FOR EMPATHY AND EDUCATION

I argue here that the gulf in *Joola*-related media represents a missed opportunity for us to humanize a tragedy that occurred beyond our doorstep. Through deeper investigation, the *Joola* wreck will enable us to learn more about Senegal, its people, and their resilience in the face of disaster, as well as the external threats they continue to face.[17] Such stories—whether in Senegal, Seoul, or Southampton—have the potential to unite us through a geography of empathy. After all, the RMS *Titanic*[18] itself was an international collaboration that evolved into one of the most popular, global narratives of our time. Constructed by Belfast-based shipbuilders, financed by American financier J. P. Morgan, and designed by British naval architect Thomas Andrews, the New York-bound *Titanic* contained passengers representing twenty-seven different nationalities (Biel 2003). While not one African was aboard *Titanic*, the Senegalese whom I interviewed had been familiar with the wreck. They knew well the scale of the disaster and in some instances, even the names of *Titanic*'s passengers. Friends in Casamance described how they had gleaned their information from books—including children's and graphic novels, mostly in French—over 180 or so of which had been published since April 1912 (Mericle 2016). And, because Cameron's film *Titanic* was released in France in 1998, just four years before the *Joola* incident, some Senegalese had actually watched the French version shortly before their own disaster unfolded. Many interviewees even mentioned being fearful of the sea and related ferry travel after watching Cameron's watershed film.[19]

Scholars and filmmakers have inarguably dug deep into *Titanic* history, and in doing so, have bypassed the anemic disaster narrative and unearthed issues of humanity, class, and race. The Atlantic Ocean is the backdrop upon which these stories unfold. It is not too late to provide an equally nuanced treatment of the *Joola* in order to shine a light on West Africa's own humanity. By excavating the events leading up to the disaster in Senegal—its absurd colonial and cartographic foundations, the MFDC war, the divide between north and south[20]—we can now better understand the conditions that gave rise to the ship's *raison d'être*. We can move beyond the casualty count and make real the everyday lives and livelihoods of Senegalese passengers. We can draw parallels between the *Titanic* and the *Joola*, and note how, for example, Ziguinchor residents boarded the ferry in pursuit of jobs and schools not found in their own backyards. We can see, then, that Senegal's shipwreck stories are not unlike those of the disenfranchised "steerage" passengers who boarded *Titanic* in pursuit of better lives in New York. In both cases, distinct geographic forces were at work, shaping everyday human movements across the Atlantic realm. Finally, the *Joola* and *Titanic* stories allow us to better visualize the inherent graft, social injustices, and environmental processes that precipitated each wreck and defined the ways in which they were remembered. We can then see that shipwrecks are not competitions or casualty counts, but stories that have the potential to link us together; they reveal the way in which our historic migration patterns converge and coalesce.

ON THE ABSENCE AND APATHY FOR AFRICA

The most obvious explanation for why, then, *Joola*-related stories have not been, and continue to remain absent in United States media and film, is that we simply do not care. Or, more fairly, we do not care *enough* to sift past the one-page headline to understand where the disaster happened and why. This is a harsh reality, but as noted at the outset it reflects our long-standing, collective attitudes toward Africa as a space devoid of detail or density (Moorman 2014). To repeat the words of Howard French, "for many Americans, the *real* Africa is a blurred concatenation of game parks, starving infants and genocidal warfare—or it's a Disney-fied cradle of civilization" (French 1994, 35). These imagined and inaccurate geographies are jumbled further by our view of Africa the *country*, rather than Africa the *continent*. The vast majority of people continue to see Africa as a homogenous and blank space on the map, ignoring its cultural, physical, political, and economic diversity (Barton 2018). Africa in fact contains 11.73 million square miles of land populated by 1.2 billion people who speak more than 200 different languages (World Bank 2020). These linguistic groups traverse eight different geographic biomes ranging from semi-arid desert to tropical forests. Yet former vice-presidential candidate Sarah Palin once called Africa a "country," and in August 2014, former senator and vice-president Joe Biden mistakenly labeled it a "nation" during the U.S.-Africa Summit. Such ignorance is universal and reaches across party lines.

In some ways, Africa is still a blurry image in the mind of many Americans—warring, impoverished, unfixable (Moorman 2014, 1). But then there are the stories of Africa the West does not hear about: urban farmers feeding their families on unclaimed plots of land, a nonprofit building mapping apps to combat election fraud and violence, the booming Nigerian "Nollywood" producing blockbuster movies (Pew Report 2010; Moorman 2014). It could be argued that the *Joola* disaster—given its sheer scale and human loss—perpetuates the myth of an "unfixable" continent replete with tragedy. Yet as this book has illustrated, in order to unravel existing stereotypes, we must "go deep" and explore the human and physical dimensions of a particular event. In other words, it is not what story we tell—or whether it upends the disaster-disease-poverty narrative of the continent—but *how* we tell it in our classrooms and communities. In the case of Africa's greatest shipwreck, I've situated the disaster within the context of a nation that was geographically, politically, and economically divided. This enables us to understand fundamental geographic principles—colonial geopolitics, rural-urban divide, cultural diversity, climate and coastal processes—that shaped the lives and livelihoods of West African people. The south's isolation from the north at large, the separatist war in Casamance, The Gambia's position in the middle of this geographic divide—all of these circumstances facilitated the *Joola* ferry line's creation. These are not causal factors, rather they help us to understand what geographers call "the why of where," or in this case why passengers boarded that overnight ferry in the first place rather than taking the long road north. Framing the story through a geographic lens enables us to enliven the event so we can then memorialize the Jola fishermen, farmers, and students who were lost, while celebrating those who survived in a spectacular show of resilience and optimism.

ON THE GEOGRAPHY OF OPTIMISM AND RESILIENCE

In 2002, Casamance resident Alpha Oumar was one of the many students who had left Ziguinchor to earn a bachelor's degree at Cheikh Anta Diop University in Dakar. In the wake of the disaster, his father told the story of how Alpha boarded the *Joola*, but was one of the 1,863 majority who did not survive when the ship sank. Alpha was a second-year geography student with the dream of becoming a university professor, his story not unlike so many others on board. While the fate of Alpha and all the victims of the *Joola* were both preventable and senseless, Murid ideology (the mystic philosophy of Sufism) provides a lens through which Senegalese view the world and make meaning out of all lived experiences, including the most tragic. Once again, some 95 percent of Senegalese are Muslim, and *rafet njort*—a central murid tenet and a word that means "beautiful optimism in the face of difficulty"—is a worldview that informs everyday life (Ngom 2016). This narrative is dedicated to Alpha Oumar and all Senegalese people in the hopes that their stories will be shared in the spirit of *rafet njort* rather than the Afro-pessimism that pervades contemporary stories about the continent. May their resilience be an inspiration to all of us.

NOTES

1. The author has presented on these research results in Mexico City, Mexico, Cairns, Australia, Denver, Colorado, Boston, Massachusetts, and Dakar, Senegal, in addition to arranging several undergraduate class presentations and university keynotes.
2. Indictment #S(9) 98 Cr. 1023, Archived March 24, 2012, at the Wayback Machine. United States District Court, Southern District of New York.
3. The idiom "the new normal"—in the sense of the new standard of baseline expectation or experience—has occurred frequently in the past fifteen years, particularly in the wake of 9/11. Books such as *The New Normal: How FDNY Firefighters Are Rising to the Challenge of Life After September 11* (2002), *Assessing the New Normal: Liberty and Security for the Post–September 11 United States* (2003), *After*

911 in the New Normal: Who Are We? Why Are We Here? Where Are We Going? (2003) underscore this fact.

4. Mayer (2009) investigates many of these societal challenges in *The Dark Side: the Inside Story of How the War on Terror Turned into a War on American Ideals,* New York, Anchor Press.

5. Families of the twenty-two French nationals who lost individuals onboard the *Joola* continue to request reparations.

6. Most of the fuel onboard the *MV Prestige* was grade #4, also known as marine fuel or surface oil; a fraction is obtained from petroleum distillation, either as a distillate or a residue. The heavy oil recovery operation took several months, and samples were collected continuously by the recovery vessels.

7. The author has used this excellent media resource in her undergraduate geography courses for ten years.

8. See PBS: Frontline, *The Lawless Sea* (2008).

9. Gross tonnage refers to the nonlinear measure of a ship's overall internal volume.

10. Demographic data on the wreck was not gathered by the Senegalese government, but by Dr. Charles Becker. The histogram was created by undergraduate research assistant Mr. Michael Chenore of UNC.

11. Most reports estimate that around 1,000 bodies remained trapped onboard the submerged *Joola* wreck (BBC 2003; Becker 2012).

12. According to the World Bank, South Korea ranks twelfth in terms of GDP per capita, whereas Senegal, one of the more robust nations in Africa, ranks 105th among all nations.

13. Director Mati Diop (*Atlantics*) is one other recent example of Senegalese cinema.

14. The Senegalese ship sailed into a storm at 11:00pm and the *Titanic* struck an iceberg at 11:40pm.

15. This footage was used, for instance, in "A Night to Remember" (1958).

16. Haidar now does most of his work on mangrove forest restoration in the Casamance region and has moved on from the Joola in terms of his primary focus (personal communication 2018).

17. The political ecology of fishing and climate change has been discussed in chapters 3 and 4.

18. The author would be remiss if she did not repeat the fact that more souls died on board the *Joola* than the *Titanic*. This is not done so for the purposes of competition, but to emphasize the sheer scale relative to its coverage and impact.

19. The author's semi-structured interviews with forty-five local families in Casamance, Senegal, July–November 2018.

20. Further discussion of these issues can be found in chapters 2–5.

References

Agence France-Presse. 2016. "Killer Hippos Spread Fear Among Fishermen in Senegal Rivers." May 26, 2018. https://www.telegraph.co.uk/news/2016/05/28/killer-hippos-spread-fear-among-fishermen-in-senegal-rivers/.

Agence France-Presse. 2018. "Senegal Vows to Hunt Down Gunmen After Massacre in Remote Forest." January 8, 2018. https://www.scmp.com/news/world/africa/article/2127270/senegal-vows-hunt-down-gunmen-after-massacre-remote-forest.

Ahmida, Ali Abdullatif (ed.). 2009. *Bridges Across the Sahara: Social, Economic and Cultural Impact of the Trans-Sahara Trade During the 19th and 20th Centuries.* Newcastle upon Tyne: Cambridge Scholars Publishing.

Aidara, Nassardine. 2003. *The Victims of the Joola Boat: The Homage of a Father* (Published in French). Dakar: Saint Paul.

Alchetron. 2018. "MV Joola Specifications." https://alchetron.com/MV-Le-Joola.

Apsaidl. 2019. "Sediou Cultural Center/AMKNA Design Studio." http://www.apsaidal.com/sedhiou-cultural-center-amkna-design-studio/.

Architectural Review. 2019. "Senegal's Peace Pavilion." https://www.architectural-review.com/competitions/competition-sedhiou-peace-pavilion-senegal/10038751.article, retrieved July 18, 2019, removed.

Archives Nationales du Sénégal. 3F/00133. 1942–44. Prison des cercles.

Austen, Ralph. 2010. *Trans-Saharan Africa in World History.* New York: Oxford University Press.

Auvray, Patrice. 2012. *Remember the Joola: Testimony of a Survivor in Memory of the 2000 Abandoned Victims* (Published in French). Paris: Vincennes Publishing.

Auvray, Patrice. 2014. *Beyond Casamance: Poems to the Forgotten of the Joola.* Dakar: Athena Press (in French).

Ayisi, F., and C. Brylla. 2013. "The Politics of Representation in the Audience Reception: Alternative Visions of Africa." *Research in African Literatures* 44(2): 125–141. Indiana University Press.

Azuma, John Alembillah. 2001. *The Legacy of Arab-Islam In Africa: A Quest for Inter-Religious Dialogue.* London: One World Publications.

Babou, Cheik Anta. 2005. "Contesting Space, Shaping Places: Making Room for the Muridiyya in Colonial Senegal, 1912–1945." *The Journal of African History* 46(3): 405–426.

Babou, Cheik Anta. 2007. "Urbanizing Mystical Islam: Making Murid Space in the Cities of Senegal." *The International Journal of African Historical Studies* 40(2): 197–223.

Barton, Karen. 2017a. "Exploring the Benefits of Field Trips in a Food Geography Course." *Journal of Geography* 116(6): 237–349.

Barton, Karen. 2017b. "The Geographical Dimensions of Africa's Greatest Shipwreck". *Focus on Geography* 60: 1–12. doi:10.21690/foge/2016.60.4.

Barton, Karen. 2018. "An African Teaching Module: Using a Shipwreck Story to Refine Students' Knowledge of Place." *The Geography Teacher* 16(2): 48–56.

Baum, Robert. 2015. *West Africa's Women of God: Aline Sitoe and the Diola Prophetic Tradition*. Bloomington: Indiana University Press.

Bauman, Zygmunt. 1992. *Mortality, Immortality and Other Life Strategies*. Stanford, CA: Stanford University Press.

Bayer Science. "Small Farms with a Large Impact." www.cropscience.bayer.com.

BBC News. 2002a. "Hundreds Lost as Senegal Ferry Sinks." *BBC News: World Edition*, September 27, 2002. http://news.bbc.co.uk/2/hi/africa/2285092.stm.

BBC News. 2002b. "Q and A: What Caused the *Joola* Ferry Disaster?" *BBC News: World Edition*, October 1, 2002. http://news.bbc.co.uk/2/hi/africa/2290490.stm.

BBC News. 2002c. "Senegal Army 'Left' Ferry Survivors." *BBC News: World Edition*, November 6, 2002. http://news.bbc.co.uk/2/hi/africa/2409087.stm.

BBC News. 2017. "Sewol Disaster Ferry Raised in South Korea After Three Years." *BBC News*, March 23, 2017.

Becker, Charles, and Idrissa Diallo. 2012. *The Sinking of the Joola Ten Years Later: Songs for the Dead* (Published in French). Paris: Editions Karthala.

Belhabib, Dyhia, Krista Greer, and Daniel Pauly. 2017. "Trends in Industrial and Artisanal Catch Per Effort in West African Fisheries." *Conservation Letters: A Journal of the Society for Conservation Biology* 11: 35–42.

Belhabib, Dyhia, Vicky Lam, and William Cheung. 2006. "Overview of West African Fisheries Under Climate Change: Impacts, Vulnerabilities, and Adaptive Responses of the Artisanal and Industrial Sectors." *Marine Policy* 9(71): 15–28.

Biel, Steven. 2012. *Down with the Old Canoe: A Cultural History of the Titanic Disaster*. New York: W.W. Norton.

Blundo, Giorgio. 2001. "Negotiating the State on a Daily Basis: Business Agents, Brokers and Touts in the Interstices of the Senegalese Administration." *Autrepart* 20: 75–90.

Boas, Morten, and Kevin Dunn (eds). 2007. *African Guerillas: Raging Against the Machine*. New York: Abe Books.

Bortei-Doku Aryeetey, E. 2002. "23 Socioeconomic Aspects of Artisanal Marine Fisheries Management in West Africa." *Large Marine Ecosystems* 11: 323–344.

Bosha, Sarah, Michelle Adeniyi, Jenna Ivan, Roya Ghiaseddin, Fabakaray Minteh, Laim Barrow, and Rex Kuye. 2007. "The Impact of the Presidential Alternative

Treatment Program on People Living with HIV and the Gambian HIV Response." *Health and Human Rights* 21(1): 239–252.

Breeden, Aurelian. 2018. "Wrath of Coastal Erosion Is Devouring a Senegal Fishing Hub." *The New York Times*, May 23, 2018. https://www.nytimes.com/, retrieved November 10, 2019.

Breetzke, G., S. Eksteen, and E. Pretorious. 2011. "Paper-Based GIS: A Practical Answer to the Implementation of GIS Education into Resource-Poor Schools in South Africa." *Journal of Geography* 110(4): 148–157.

Brooks, George. 1977. "Peanuts and Colonialism: Consequences of the Commercialization of Peanuts in West Africa: 1830–1970." *The Journal of African History* 16(1): 29–54.

Burke, Jason. 2018. "Tanzania Ferry Disaster: Divers Pull Survivor from Capsized Ship". *The Guardian*, September 23, 2018.

Capet, X., P. Estrade, E. Machu, S. Ndoye, J. Grelet, A. Lazar, L. Marie, D. Dausse, and P. Brehemer. 2017. "On the Dynamics of the Southern Senegal Upwelling Center: Observed Variability from Synoptic to Superinertial Scales." *Journal of Physical Oceanography*, January 2017. doi:10.1175/JPO-D-15-0247.1

CNN. 2002a. "Hundreds Missing as Ferry Sinks." September 27, 2002. www.cnnworld.com, retrieved June 10, 2019.

CNN. 2002b. "Senegal Ferry Disaster Tops 300." September 28, 2002. www.cnnworld.com, retrieved June 10, 2019.

Cochrane, Laura. 2016. "Rebuilding Agriculture and the Environment in Senegal." *Anthropology Now* 8(1): 47–57.

"Commission d'Enquete Technique sur les Causes du Naufrage du 'Joola', Submitted to the President of the Republic on November 4, 2002, Dakar." http://www.bea-mer.developpement-durable.gouv.fr/IMG/pdf/RET_JOOLA_2002_Site.pdf, retrieved October 7, 2019.

Cormier-Salem, M. C. 1994. "Environmental Changes, Agricultural Development, and Small Scale Fishing Development in the Casamance Region, Senegal." *Ocean and Coastal Management* 24(2): 109–124.

Coulon, Christian. 1999. "The Grand Magal in Touba: A Religious Festival of the Mouride Brotherhood of Senegal." *African Affairs* 98(391): 195–210.

Crowder, Michael. 1968. *West Africa Under Colonial Rule*. Chicago: Northwestern University Press.

Cruise Ship Sinking. 2002. "MV Le Joola Capsizes 25-September 2002." September 25, 2002. www.cruiseshipsinking.com, retrieved February 15, 2020.

De Jong, F., and G. Gasser. 2005. "Contested Casamance: Introduction." *Canadian Journal of African Studies* 392(2): 213–229.

De Jong, Ferdinand. 2005. "A Joking Nation: Conflict Resolution in Senegal." *Canadian Journal of African Studies* 39(2): 389–413.

De Jong, Steffi. 2018. *The Witness as Object: Video Testimony in Memorial Museums.* New York: Berghahn Books.

Deets, Mark. 2016. "Grown-Ups on White Plastic Chairs: Soccer and Separatism in Senegal, 1969–2012." *History in Africa* 43(1): 347–374.

Diawara, Manthia. 2000. *In Search of Africa*. Cambridge, MA: Harvard University Press.

Dinar, A., R. Hassan, R. Mendelsohn, and J. Benhin (eds). 2008. *Climate Change and Agriculture in Africa: Impact Assessment and Adaptation Strategies*. London: Earthscan.

Diome, Fatou. 2019. *Les Veilleurs de Sangomar* (French Edition). Paris: Albin Michel.

Direction de l'Administration Pénitentiaire (DAP). 1968. "Enquête sur la criminalité au Sénégal année 1967." Dakar, Sénégal.

Documentaire. 2015. "Rideau Rouge sur la Naufrage du Joola." https://www.youtube.com/watch?v=-3pMgLCvsAY&feature=youtu.be.

Dodd, Mark. 2015. *Ethiopia Rising*. Journeyman Pictures, distributed September 15, 2015.

Doumbouya, A., O. Camara, J. Mamie, J. Intchama, A. Jarra, S. Ceesay, A. Gueye, D. Ndiaye, E. Beibou, A. Padilla, and D. Belhabib. 2017. "Assessing the Effectiveness of Monitoring Control and Surveillance of Illegal Fishing: The Case of West Africa." *Frontiers in Marine Science* 4(5): 1–10.

DuBois, Carolyn, and Christos Zografos. 2012. "Conflicts at Sea Between Artisanal and Industrial Fishers: Inter-Sectoral Interactions and Dispute Resolutions in Senegal." *Marine Policy* 36: 1211–1220.

Dwyer, Maggie. 2017. "Fragmented Forces: The Development of the Gambian Military." *African Security Review* 26(4): 362–377.

Eliade, Mircea. 1987. *The Sacred and the Profane: The Nature of Religion*. San Diego: Harcourt Brace Jovanovich.

Eltringham, Nigel (ed.). 2013. *Framing Africa: Portrayals of a Continent in Contemporary Mainstream Cinema*. New York: Berghahn Books.

Endangered Archives Program. 2019. "Digital Preservation of Mandinka Ajami Manuscripts, Casamance, Senegal." https://eap.bl.uk/project/EAP1042, retrieved January 6, 2020.

Environmental Justice Atlas. 2018. "Illegal Logging and Trafficking of Teakwood in the Forest of Bayottes, Senegal." May 16, 2018. https://ejatlas.org/conflict/senegal.

ESRI. 2016. "Geography, Class and Fate: Passengers Onboard the Titanic." *ESRI Story Maps*. http://storymaps.esri.com/stories/titanic/, retrieved December 15, 2020.

Eusébio, Dasylva. 2005. *Le Naufrage du Joola ou Devoir de Mémoire*. Ziguinchor: Coédition NENA/Editions Salamata.

Evans, Martin. 2000. "Briefing: Senegal: Wade and the Casamance Dossier." *African Affairs* 99: 649–658.

Evans, Martin. 2002. "The Casamance Conflict; Out of Sight, Out of Mind?" *Humanitarian Practice Network*, April 2002.

Evans, Martin. 2005. "Insecurity or Isolation? Natural Resources and Livelihoods in Lower Casamance." *Canadian Journal of African Studies* 9: 282–312.

Evans, Martin. 2007. "The Suffering Is Too Great: Urban Displaced Persons in the Casamance Conflict, Senegal." *Journal of Refugee Studies* 20(1): 60–85.

Evans, Martin. 2009. "Flexibility in Return, Reconstruction and Livelihoods in Displaced Villages in Casamance, Senegal." *GeoJournal* 74: 507–524.

Ewens, Graeme. 2002. "The Tragedy of Le Joola: Graeme Ewens Reports on the Possible Causes of Africa's Worst Ever Shipwreck and the Political Fallout in Senegal." *African Business*, December 2002, pp. 44–45.

Fare, Yohann, Marc Dufumier, Myriam Loloum, Fanny Miss, Alassane Pouye, Ahmat Khastalani, and Adama Fall. 2017. "Analysis and Diagnosis of the Agrarian System in the Niayes Region, Northwestern Senegal." *Agriculture* 7(7): 59. https://www.mdpi.com/2077-0472/7/7/59/htm.

Fisher, Max. 2012. "The Dividing of a Continent: Africa's Separatist Problem." September 20, 2012. www.theatlantic.com, retrieved February 8, 2020.

Food and Agricultural Organization. 2008. "Senegal Fishery Production System Report: Artisanal Fisheries Subsector." wwww.firms.fao.org/firms/fishery/473/en.

Food and Agricultural Organization. 2015a. "Country Fact Sheet on Food and Agricultural Policy Trends: Republic of Senegal." http://www.fao.org/3/a-i4841e.pdf.

Food and Agricultural Organization. 2015b. "Farmer Field Schools: A Participatory Approach to Capacity Building for Efficient, Sustainable and Inclusive Food Production." www.fao.org/senegal/actualites/detail-events/en/c/358275/.

Foote, Kenneth. 2003. *Shadowed Ground: American Landscapes of Violence and Tragedy*. Austin: University of Texas Press.

Foucher, Vincent. 2011. "On the Matter (and Materiality) of the Nation: Interpreting Casamance's Unresolved Separatist Struggle." *Studies in Ethnicity and Nationalism* 11(1): 82–103.

French, Howard. 1994. "The World: An Ignorance of Africa as Vast as the Continent." November 20, 1994. Section 4, page 3. https://www.nytimes.com/1994/11/20/weekinreview/the-world-an-ignorance-of-africa-as-vast-as-the-continent.html, retrieved February 15, 2020.

Gansler, Katri. 2018. "Senegal Cow Power." *Akzente: the GIZ Magazine*, January 2018. https://akzente.giz.de/en/artikel/cow-power.

Gashaw, Tasew. 2017. "Colonial Borders in Africa: Improper Design and Its Impact on African Borderland Communities." November 17, 2017. https://africaupclose.wilsoncenter.org/colonial-borders-in-africa-improper-design-and-its-impact-on-african-borderland-communities/, retrieved February 8, 2020.

Gershmel, Phil. 2018. "Lesson Plan: Trade in the Ancient World—Evidence from a Shipwreck." *The Geography Teacher* 15(1): 19–20.

Godard, Xavier, and Pierre Teurnier. 1992. *Urban Transport in Africa at the Time of Structural Adjustment*. Paris: Karthala.

Gordon, David, and Shepard Krech III (eds). 2012. *Indigenous Knowledge and the Environment in Africa and North Africa*. Athens: Ohio University Press.

Gounongbe, Ari. 2003. *In the Storm of the Joola* (Published in French). Saint-Louis, Senegal: Xamal Publishing.

Hayden, Brian. 2009. "Funerals as Feasts: Why Are They So Important?" *Cambridge Archaeological Journal* 19(1): 29–52.

Heil, Tilmann. 2014. "Are Neighbors Alike? Practices of Conviviality in Catalonia and Casamance." *European Journal of Cultural Studies* 17(4): 452–470.

Hockels, L. (2002). “Maritime Report. Ed Paul Redfern in The East African Report.” www.nationaudio.com/news/eastafrican/21102002/maritime/maritime18.html.

International Federation of the Red Cross. 2002. “Maritime Safety Ignored in Senegal.” www.irfc.com.

International Federation of the Red Cross. 2003. “Senegal Ferry Disaster Reveals Need for Psychological Support.” March 3, 2003. https://www.ifrc.org/en/news-and-media/news-stories/africa/senegal/senegal-ferry-disaster-reveals-need-for-psychological-support/.

Jacobs, Andrew. 2017. “China’s Appetite Pushes Fisheries to the Brink in Africa.” *New York Times*. https://www.nytimes.com/2017/04/30/world/asia/chinas-appetite-pushes-fisheries-to-the-brink.html.

Jar Torre, Luis. 2003. “An African Massacre: The Wreck of the Joola.” *Marina General Magazine*, November 2003. http://www.grijalvo.com/Batracius/Batracius_Joola.htm.

Jindra, Michael. 2005. “Christianity and the Proliferation of Ancestors: Changes in Hierarchy and Mortuary Ritual in the Cameroon Grass-fields.” *Africa* 75(3): 356–377.

Jones, Hillary. 2013. *The Metis of Senegal: Urban Life and Politics in French West Africa.* Bloomington: Indian University Press.

Jullien, Maud. 2012. “Africa’s Titanic: Seeking Justice a Decade After Joola.” *BBC News*, September 26, 2012. https://www.bbc.com/news/world-africa-19717929, retrieved December 1, 2019.

Kama, Khady. 2010. “La Commune de Joal Fadiouth: Etude des Inondations.” In *Mémoire de Maitrise*. Dakar: Université Cheikh Anta Diop de Dakar.

Keaten, James. 2002. “Divers Describe Ferry Disaster Scene.” *Associated Press News*, September 30, 2002. https://apnews.com/a406d5ec772133d2d3bbf05299921f8d.

Kim, Suk Kyoon. 2017. “The Sewol Ferry Disaster in Korea and Maritime Safety Management.” In *Maritime Disputes in Northeast Asia*. Leiden: Brill, pp. 183–199.

Kimble, George. 1951. “Africa Today: The Lifting Darkness.” *Reporter*, May 15, 1951, p. 157.

Kingston, Kato Gogo. 2011. “The Impacts of the World Bank and IMF Structural Adjustment Programmes on Africa: A Case Study of Cote D’Ivoire, Senegal, Uganda and Zimbabwe.” *Sacha Journal of Policy and Strategic Studies* 1(2): 110–130.

Klein, Martin. 1968. *Islam and Imperialism in Senegal: Sine-Saloum*. Stanford: Stanford University Press.

Konate, Dior. 2003. “Ultimate Exclusion: Imprisoned Women in Sénégal.” In Florence Bernault (ed.) *A History of Prison and Confinement in Africa*. Portsmouth, NH: Heinemann, pp. 155–164.

Lagos This Day. 2002. “Senegal Report” Ed Barbara Slaughter. “Senegal Ferry Disaster Kills Close to a Thousand Passengers.” October 2, 2002.

Lambert, Michael. 2002. *Longing for Exile: Migration and the Making of a Translocal Community in Senegal, West Africa.* Portsmouth: Heinemann.

Langewiesche, Katrin.2011. "Funerals and Religious Pluralism in Burkina Faso." In Jindra Michael and Noret Joel (eds) *Funerals in Africa: Explorations of a Social Phenomenon*. New York, NY: Berghahn Books, pp. 130–153.

Lavery, Sean. 2019. "The Sinking of the M.V. Sewol and the Confusion of Disasters." *The New Yorker*, April 9, 2019.

Leary, Fay. 1970. "Islam, Politics and Colonialism: A Political History of Islam in the Casamance Region of Senegal (1850–1914)." PhD Dissertation, Northwestern University.

Leary, Fay. 1971. "The Role of the Mandinka in the Islamization of the Casamance, 1850–1901." In C. T. Hodge (ed.) *Papers on the Manding*. Bloomington, IN: Indiana University, pp. 227–248.

Lee, Su-Huyn, and Choe Sang-Hun. 2014. "Students Among Hundreds Missing as South Korean Ferry Sinks." *The New York Times*, April 16, 2014.

Levitt, Peggy, and Nina Glick Schiller. 2004. "Conceptualizing Simultaneity: A Transnational Social Field Perspective on Society." *International Migration Review* 38: 1002–1039.

Levtzion, Nehemiah, and Randall Pouwels. 2000. *The History of Islam in Africa*. Athens: Ohio University Press.

Lewis, Ioan.1966. *Islam in Tropical Africa*. Oxford: Oxford University Press.

Liberto, Tom. 2015. "Wait, a Hurricane Formed Where in the Atlantic?" September 3, 2015. https://www.climate.gov/, retrieved January 15, 2020.

Linares, O. F. 1992. *Power, Prayer and Production: The Jola of Casamance, Senegal*. Cambridge: Cambridge University Press.

Linares, Olga. 1996. "Cultivating Biological and Cultural Diversity: Urban Farming in Casamance, Senegal." *Africa: Journal of the International African Institute* 66(1): 104–121.

Linares, Olga. 2003. "Going to the City . . . and Coming Back? Turnaround Migration Among the Jola of Senegal." *Africa* 73: 113–132.

Loimeier, Roman. 2013. "The Sahara as Connective Space." In Roman Loimeier (ed.) *Muslim Societies in Africa: A Historical Anthropology*. Bloomington: Indiana University Press, pp. 54–76.

Lombard, Jérôme. 2003. "Senegal: Drifts from the Transport System to the *Joola* Disaster." *Contemporary Africa* 3(207): 165–184.

Mahling Clark, K. 2011. "Ripe or Rotting: Civil Society in the Casamance Conflict." *African Conflict and Peace Building Review* 12: 153–172.

Marut, Jean-Claude. 2002. "Is the Casamance Problem Soluble in the Nation-State?" In Momar-Coumba Diop (ed.) *Contemporary Senegal*. Paris: Karthala, pp. 426–458.

Mendelhson, Daniel. 2012. "Unsinkable: Why We Can't Let Go of the Titanic." *The New Yorker*, April 9, 2012. https://www.newyorker.com/magazine/2012/04/16/unsinkable-3.

Mericle, Kate. 2016. "Content Analysis of Media Articles Related to the Joola Ferry Wreck of 2012." Unpublished Undergraduate Research Project, University of Northern Colorado.

Ministry of Economy and Finance. 2001. *Poverty Reduction Strategy Paper*. Dakar.

Ministry of Economy, Finance and Planning. 1999. *Economic and Social Situation of Senegal* (1998 Edition). Dakar: Directorate of Forecasting and Statistics.

Ministry of Equipment and Transport. 1998. *Memento on land transport in Senegal*. Dakar: Directorate of Land Transport.

Ministry of Equipment and Transport. 2002. *Memento of Land Transport of Senegal*. Dakar: Directorate of Land Transport.

Mohamed, Hamza. 2017. "Remembering the Gambia's Notorious Mile 2 Prison." *Al Jazeera*, March 29, 2017. https://www.aljazeera.com/indepth/features/2017/03/remembering-gambia-notorious-mile-2-prison-170322073015550.html, retrieved February 8, 2020.

Monmonier, Mark. 2018. *How to Lie with Maps*. Chicago: University of Chicago Press.

Moseley, William. 2001. "Computer Assisted Comprehension of Distant Worlds: Understanding Hunger Dynamics in Africa." *Journal of Geography* 100(1): 32–45.

National Centers for Environmental Information. 2002. "Hurricanes and Tropical Storms: 2002." In *State of the Climate: Hurricanes and Tropical Storms*. NOAA.

National Public Radio. 2010. "Africa: For Many in Senegal, Statue Is a Monumental Failure." January 5, 2010. https://www.npr.org/templates/story/story.php?storyId=122220923.

Neiland, Arthur, and Christophe Bene (eds). 2004. *Poverty and Small Scale Fisheries in West Africa*. Netherlands: Springer.

New York Times. 1964. "Peanut Harvest Rises in Senegal: Nation Is Seeking Diversity as Subsidy Nears End." January 20, 1964, p. 76. www.newyorktimes.com.

New York Times. 2013. "Deaths at Sea: Grim Numbers." February 20, 2003. https://www.nytimes.com/2003/02/20/international/africa/deaths-at-sea-grim-numbers.html?searchResultPosition=2.

New York Times. 2019. "Millennials Make Farming Sexy in Africa, Where Tilling the Soil Once Meant Shame." May 27, 2019. https://www.nytimes.com/2019/05/27/world/africa/farming-millennials.html.

News 24. 2018. "Three Gambians Arrested Over Senegalese Forest Guard Murder." April 10, 2018. www.news24.com, retrieved February 18, 2020.

Ngom, Fallou. 2010. "Ajami Scripts in the Senegalese Speech Community." *Journal of Arabic & Islamic Studies* 10(1): 1–23.

Ngom, Fallou. 2016. *Muslims Beyond the Arab World: The Odyssey of Ajami and the Muriddya*. New York: Oxford University Press.

Nietschmann, Bernard. 1994. "The Fourth World: Nations Versus States." In George J. Demko and William Wood (eds). *Reordering the World: Geopolitical Perspectives on the Twenty-First Century*. Boulder, CO: Westview Press, pp. 225–242.

Nietschmann, Bernard. 1995a. "Coral Reef Conservation by Local Communities." In S. F. Stevens (ed.) *Conservation Through Cultural Survival: Indigenous Peoples and Protected Areas*. Washington, DC: Island Press.

Nietschmann, Bernard. 1995b. "Defending the Miskito Reefs with Maps and GPS: Mapping with Sail, SCUBA, and Satellite." *Cultural Survival Quarterly* 18(4): 34–37.

NOAA. 2020. "Climate Prediction Center for West Africa." *National Weather Service*. https://www.cpc.ncep.noaa.gov/products/global_monitoring/precipitation/wafrica_1yrprec.shtml.

Nugent, Paul. 2007. "Cyclical History in the Gambia/Casamance Borderlands: Refuge, Settlement and Islam from c. 1880 to the Present." *The Journal of African History* 48(20): 221–243.

Nyssen, Jan, Amaury Frankl, Mitiku Haile, Hans Hurni, Katrien Descheemaeker, Donald Crummey, Alfons Ritler, Brigitte Portner, Bernhard Nievergelt, Jan Moeyersons, Neil Munro, Jozef Deckers, Billi Paolo, and Jean Poesen. 2014. "Environmental Conditions and Human Drivers for Changes to North Ethiopian Mountain Landscapes Over 145 Years." *Science of the Total Environment* 485–486: 164–179.

Obasanjo, Olusegun. 2012. "How Africa Could Feed the World." November 6, 2012. https://globalpublicsquare.blogs.cnn.com/2012/11/06/how-africa-could-feed-the-world/, retrieved December 1, 2019.

Oduntan, Gbenga. 2015. "Africa's Border Disputes Are Set to Rise, But There Are Ways to Stop Them." July 14, 2015. https://theconversation.com/africas-border-disputes-are-set-to-rise-but-there-are-ways-to-stop-them-44264, retrieved February 8, 2020.

Oguh, Chibbuike. 2015. "The Representation of Africa in Western Media: Still a 21st Century Problem." Master's Thesis, Edinburgh Napier University.

Ogundayo, BioDun, and Julius Adekunle (eds). 2019. *African Sacred Spaces: Culture, History and Change*. Lanham: Rowman and Littlefield.

Okieimen, F. E., and C. O. Eromosele. 1999. "Fatty Acid Composition of the Seed Oil of Khaya Senegalensis." *Bioresource Technology* 69(3): 279–280. doi:10.1016/S0960-8524(98)00190-4.

One Campaign. 2014. "Africa's Biggest Music Stars & One.org Launch Cocoa na Chocolate to Revolutionize Agriculture." March 30, 2014. www.one.org/africa, retrieved January 15, 2020.

Onoma, Ato Kwamena. 2018. "The Grave Preferences of Mourides in Senegal: Migration, Belonging, and Rootedness." *Africa Spectrum* 53(3): 65–88.

Open Water Swimming Association. 2020. "J'ai Fait Dakar-Gorée." https://openwaterswimming.com/2020/02/jai-fait-dakar-goree/.

Osunde, S. 1996. "The Politics and Social Influence of News-Making: American Media Coverage of Africa." PhD Dissertation, Howard University, Washington, D.C.

Overseas Development Institute. 2014. "West Africa's Missing Fish: The Impacts of Illegal, Unreported, and Unregulated Fishing and Under-Reporting Catches by Foreign Fleets." https://www.odi.org/sites/odi.org.uk/files/resource-documents/10665.pdf.

Panos IPAO. 2004a. *The Sinking of the Joola: Citizen Responsibility of the Senegalese Media*. Dakar: Panos Institute for West Africa.

Panos IPAO. 2004b. *The Sinking of the Joola: Media and Civil Society Facing the Tragedy*. Dakar: Panos Institute for West Africa.

Park, Chris. 1994. *Sacred Worlds: An Introduction to Geography and Religion*. New York: Taylor and Francis.

Park, Chris. 2004. "Religion and Geography." In John Hinnels (ed.) *Routledge Companion to the Study of Religion*. London: Routledge, pp. 1–29.

Parker, John, and Richard Reid. 2013. *The Oxford Handbook of Modern African History*. Oxford: Oxford University Press.

Pasche, Richard, Miles Lawrence, Lixion Avila, John Beven, James Franklin, and Stacy Stewart. 2004. "Atlantic Hurricane Season of 2002." *Tropical Prediction Center, National Hurricane Center*, July, pp. 1829–1859.

Pew Forum. 2010. *Tolerance and Tension: Islam and Christianity in Sub-Saharan Africa*. New York: Pew Forum on Religion and Public Life.

Pew Research Center. 2015. "Cell Phones in Africa: Communication Lifeline." April 15, 2015. https://www.pewresearch.org/global/2015/04/15/cell-phones-in-africa-communication-lifeline/, retrieved January 23, 2020.

Plazier, Wim. 2016. "How Africa Can Feed the World." January 22, 2016. www.weforum.org, retrieved June 1, 2019.

Poncian, Japhace. 2015. "The Persistence of Western Negative Perceptions About Africa: Factoring in the Role of Africans." *Journal of African Studies and Development* 7(3): 72–80.

Quinn, Charlotte. 1971. "A Nineteenth Century Fulbe State." *The Journal of African History* 12: 427–440.

Quinn, Charlotte. 1972. *Mandingo Kingdoms of the Senegambia: Traditionalism, Islam, and European Expansion*. London: Longman.

Randall, Sara, and Ernestina Coast. 2016. "The Quality of Demographic Data on Older Africans." *Demographic Research* 34(5): 143–174.

Reboussin, D. 1995. *From Affiniam-Boutem to Dakar: Migration from the Casamance, Life in the Urban Environment of Dakar, and the Resulting Evolutionary Changes in Local Diola Organizations*. Gainesville, FL: University of Florida.

Reese, Scott. 2014. "Islam in Africa/Africans and Islam." *Journal of African History* 55: 17–26. Cambridge University Press.

Republic of Senegal. 2002. Technical Commission of Inquiry into the Causes of the Sinking of the Joola Investigation Report, Dakar, 4 November 2002, pp. 1–86 (Commission d'enquête technique sur les causes du naufrage du Joola Rapport d'enquête Dakar, le 04 novembre 2002).

Riccio, B. 2005. "Talkin' About Migration: Some Ethnographic Notes on the Ambivalent Representation of Migrants in Contemporary Senegal." *Stichproben: Wiener Zeitschrift für kritische Afrikastudien* 5: 99–118.

Roche, Christian. 1985. *Histoire de la Casamance: Conquête et résistance, 1850–1920*. Paris: Karthala.

Ross, Eric. 2011. "A Historical Geography of the Trans-Saharan Trade (Map)." In Graziano Krätli and Ghislaine Lydon (eds) *The Trans-Saharan Book Trade: Manuscript Culture, Arabic Literacy and Intellectual History in Muslim Africa*. Leiden: Brill, pp. 1–34.

Rothe, Dawn, Stephen Muzzatti, and Christopher Mullins. 2006. "Crime on the High Seas: Crimes of Globalization and the Sinking of the Senegalese Ferry Le Joola." *Critical Criminology* 14: 159–180.

Saine, Abdoulaye. 2000. "The Gambia's Foreign Policy Since the Coup, 1994–1999." *Commonwealth & Comparative Politics* 38(2): 73–88.

Sall, Babacar. 2003. *Sea Salt: Poems for Shipwrecked and Survivors of the Joola.* Paris: L'Harmattan (in French).

Sang-Hun, Choe. 2019. "An Overloaded Ferry Flipped and Drowned Hundreds of School Children. Could It Happen Again?" *The New York Times*, June 10, 2019. https://www.nytimes.com/2019/06/10/world/asia/sewol-ferry-accident.html.

Sene, Ousmane. 2016. "The Sereer." *West African Research Center Lecture*, May 23, 2016.

Senegal Data Portal. 2020. "Senegal Census Data." https://senegal.opendataforafrica.org/gkdvujc/senegal-census-data, retrieved February 8, 2020.

Senghor, Diana. 2004. *The Joola Shipwreck: Media and Civil Society Facing the Drama* (Published in French). Dakar: Institute Panes Afrique de l'Ouest.

Sengupta, Somini. 2003. "African Town's Pride and Joy Vanished When Ferry Sank." *The New York Times*, February 20, 2003. https://www.nytimes.com/2003/02/20/world/african-town-s-pride-and-joy-vanished-when-ferry-sank.html?searchResultPosition=1, retrieved January 2, 2020.

Sieff, Kevin. 2020. "A Quest to Find the African Slave Ships That Sank in the Atlantic." http://digitaledition.chicagotribune.com/tribune/article_popover.aspx?guid=6ecda93d-517a-448d-af44-3217ed9c0d8a.

Sodaro, Amy. 2018. *Exhibiting Atrocity: Memorial Museums and the Politics of the Past.* New York: Rutgers University Press.

St Jacques, E. 2009. "Economic Mobility and the Transnational Practices of West Africans in Catalonia, Spain. Graduate School." PhD Dissertation, University of Florida.

Stump, Roger. 2008. *The Geography of Religion: Faith, Place, and Space*. Lanham: Rowman and Littlefield Publishers.

Summers, Hannah. 2019. "Chinese Fishmeal Plants Leave Fishermen in the Gambia at Sea." March 20, 2019. https://www.theguardian.com/global-development/2019/mar/20/chinese-fishmeal-plants-leave-fishermen-gambia-all-at-sea.

The New Humanitarian. 2018. "Senegal Families Demand Justice for Joola Ferry Deaths." September 19, 2018. http://www.thenewhumanitarian.org/report/80500/senegal-families-demand-justice-joola-ferry-deaths, retrieved January 15, 2020.

Theobald, Anne. 2015. "Successful or Failed Rebelling? The Casamance Perspective from a Framing Perspective." *Civil Wars* 17(2): 181–200.

Thiao, Djiga, Julie Leport, Baba Ndiaye, and Adama Mbaye. 2018. "Need for Adaptive Solutions to Food Vulnerability Induced by Food Scarcity and Unaffordability in Senegal." *Aquatic Living Resources* 31(25): 1–9. doi:10.1051/alr/2018009.

Thiaw, Ibrahima. 2003. "Archaeology and the Public in Senegal: Reflections on Doing Fieldwork at Home." *Journal of African Archaeology* 1(2): 27–35.

Thiaw, Ibrahima. 2011. "Slaves Without Shackles: An Archaeology of Everyday Life on Goree Island." In P. Lane and K. MacDonald (eds) *Comparative Dimensions of Slavery in Africa: Archaelogy and Memory, Proceedings of the British Academy 168.* Oxford: Oxford University Press, pp. 147–165.

Thomas, Jean-Raymond, and Michel Tricot. 2008a. "Sinking of the Senegalese Ferry Joola." *Expert Report for the High Court of Evry.*

Thomas, Jean Raymond, and Michel Tricot. 2008b. "Tribunal de Grande Instance d'Evry." *Naufrage du Transbordeur Senegalais: Le Joola*, pp. 1–265.

Thomas, Louis-Vincent. 1982. *La Mort Africaine: Idéologies Funéraires en Afrique Noire*. Paris: Payot.

Toliver-Diallo, Wilmetta J. 2005. "The Woman Who Was More Than a Man: Making Aline Sitoe Diatta into a National Heroine in Senegal." *Canadian Journal of African Studies* 39(2): 338–360.

Toure, Moussa. 2013. *The Pirogue*. Dakar: Orange Studio.

Transparency International. 2012. "Ending Impunity in Senegal." November 12, 2012. https://www.transparency.org/news/feature/ending_impunity_in_senegal.

Transparency International. 2015. "People and Corruption: Africa Survey Global Corruption Barometer." www.transparency.org/whatwedo/publication/people_and_corruption_africa_survey_2015.

Tuan, Yi-Fu. 2001. *Space and Place*. Minneapolis: University of Minnesota Press.

United Nations Cartographic Profile. 2004. "Senegal Map: Department of Peacekeeping Operations." http://www.un.org/Depts/Cartographic/map/profile/senegal.pdf.

United Nations Development Programme. 1999. "For Better or Worse: The Euro and CFA Franc." *African Recovery On-Line*. http://www.un.org/ecosocdevgeninfo/afrec/subjindx/124euro3.htm.

United Nations Development Programme. 2000. "Senegal." October 2003. www.un.org.

United Nations Development Programme. 2003. *Press Release*. Senegal.

United Nations Educational, Scientific and Cultural Organization. 2009. "Literacy and Non-Formal Education." www.unesco.org/new/en/dakar/education/literacy.

United States Agency for International Development. 2018. "Senegal Educational Fact Sheet." Updated May 31, 2018. https://www.usaid.gov/senegal/newsroom/fact-sheets/senegal-education-fact-sheet.

University of Assane Seck Ziguinchor. 2020. "History." https://uasz.sn/presentation/historique/.

Verschatse, Alain. 2004. *Le Joola: A Matter of State*. Lyon: Le Croquant.

Villalón, Leonardo. 2004. "Islamism in West Africa: Sénégal." *African Studies Association* 47: 61–72.

VOA News. 2009. "Senegal President Dismisses Prime Minister." October 30, 2009. www.voanews.com, retrieved February 1, 2019. https://www.voanews.com/archive/senegal-president-dismisses-prime-minister-2002-11-04.

Walker, Peter. 2010. "Senegalese President Unveils African Renaissance Statue." April 4, 2010. https://www.theguardian.com/world/2010/apr/04/senegal-african-renaissance-statue.

Walsh, Niall Patrick. 2019. "Winning Designs for Senegal Peace Pavilion, Judged by Kengo Kuma." June 11, 2019. https://www.archdaily.com/918845/winning-designs-for-senegal-peace-pavilion-judged-by-kengo-kuma, retrieved December 1, 2020.

Wanyoni, Innocent, Andrew Wamuota, Simeon Mesaki, and Almeida Guisssamulo. 2016. "Artisanal Fisher Migration Patterns in Coastal East Africa." *Ocean and Coastal Management* 119: 93–108. doi:10.1016/j.ocecoaman.2015.09.00.

Weissman, S. R. 1990. "Structural Adjustment in Africa: Insights from the Experiences of Ghana and Senegal." *World Development* 18(12): 1621–1635.

Wiley, Pat. 2012. *The Sinking of MV Le Joola: Africa's Titanic.* New York: First Steps Publishing.

Williams, David. 2012. *In Titanic's Shadow: The World's Worst Merchant Ship Disasters.* Gloucestershire: History Press.

Wilmetta, J. Toliver-Diallo. 2005. "The Woman Who Was More Than a Man: Making Aline Sitoe Diatta into a National Heroine in Senegal." *Canadian Journal of African Studies* 39(2): 340–362.

World Bank. 2016. "A Message from Senegalese Youth Farmers: Agriculture Is Full of Opportunities for Success." November 1, 2016. https://www.worldbank.org/en/news/feature/2016/11/01/a-message-from-senegalese-youth-farmers-agriculture-is-full-of-opportunities-for-success.

World Bank. 2019. "Employment in Agriculture." https://data.worldbank.org/indicator/SL.AGR.EMPL.ZS?end=2018&locations=SN&start=1991&view=chart, retrieved September 2019.

World Bank. 2020. "Life Expectancy by Birth, Senegal." https://data.worldbank.org/indicator/SP.DYN.LE00.MA.IN?locations=SN.

World Food Programme. 2011. *Climate Risk and Food Security in Senegal: Analysis of Climate Impacts on Food Security and Livelihoods.* Senegal: Dakar WFP Office, pp. 1–48. https://documents.wfp.org/stellent/groups/public/documents/newsroom/wfp269381.pdf.

World Resources Institute. 2015. "How Ethiopia Went from Famine Cross to Green Revolution." July 28, 2015. https://www.wri.org/blog/2015/07/how-ethiopia-went-famine-crisis-green-revolution.

Yeung, Peter. 2019. "Climate Change Is Already Battering This West African City." April 8, 2019. https://www.citylab.com/environment/2019/04/climate-change-rising-sea-levels-west-africa-coastal-cities/586651/.

Appendix

Adapted from: Le naufrage du Joola dix ans après: Chants pour les morts et les survivants*

This list is a tribute to all the victims of the shipwreck. It was established thanks to the considerable and voluntary work of several friends and the assistance of all those—individuals, services or associations—who collected and communicated their partial lists. The general file from which the following list is extracted was created for the families of victims, to help create a complete list of victims. It was prepared at the request of the Collectif des Familles de Victimes du Joola and the members of its Bureau, initially chaired by Guy Aris Ndouye, then by Idrissa Diallo. It was developed with the help of members of the "Identification of beneficiaries" group of the Psycho-medical Support Unit, set up after the sinking. All the regional Associations of Victims' Families and especially the Ziguinchor Community Association have sent us their censuses. We have also integrated information from various other lists, official or anonymous, as well as individual information provided by relatives and friends of victims or by press articles. About fifty lists were collected and compared to create a complete file (Becker February 20, 2020, personal communication).

Martha Abeudje, 31
Thomas Abeudje, 52
Souleymane Aidara, 17
Marie A. Aluidisse Albis, 16
Yves Alvaris
Fatoumata Kiné Anne, 50
Lisette Rike Arendshorst, 32
Roel Hank Arendshorst, 34
Frank Assarel Azare, 42
Albertine Henriette Assine, 32
Paul Christophe Assine, 32
Abdoulaye Ba, 23
Abdoulaye Ba, 62
Abdourahmane Ba, 49
Adama Ba, 14
Adama Ba, 51
Alassane Ba, 22
Amadou Diouldé Ba, 24
Amady A. Ba
Ameth Ba, 8

Babacar Boubacar Ba, 37
Bahab Ba
Boubacar Ba, 22
Daouda Ba
Diouldé Ba, 21
Faly Ba, 15
Fatimata Ba, 16
Fatou Ba, 9
Fatou Ba, 47
Ibrahima Ba, 21
Idrissa Ba, 38
Khady Khadidiatou Ba, 21
Mamadou Ba
Mame Fatou Yacine Ba, 4
Mbarka Ba
Moussa Ba, 11
Oumar Ba
Seydou Ba, 26
Souleymane Ba, 40
Yaya Ba, 23
Younouss Mohamed Lamine Ba, 11
Charles Badiane, 56
Diénaba Badiane, 23
Gaston Pascal Badiane, 26
Joseph Badiane, 25
Maciré Badiane, 32
Mamadou Badiane
Pascal Badiane, 32
Boubacar Badiane
Mbassine Badiane
Fanta Ciré Badio, 47
Abdou Badji
Adama Badji
Adiaratou Badji, 56
Aïssatou Badji, 36
Amadou Moustapha Badji, 24
Awa Badji, 36
Awa Badji, 39
Boubacar Badji
Cheikh Atab Badji, 24
Denise Anastancia Badji, 21
Emile Badji, 17
Gnima Niama Badji, 34
Idrissa Badji, 21
Innocence Inosencio Badji, 59
Jules Olympio Badji, 27
Lansana Badji
Malamine Badji
Mamadou Badji, 24
Marthe Gaédick Chérita Badji, 3
Mélanie Badji, 31
Noellie Badji, 5
Ousmane Badji, 35
Papa Ch. Badji
Rosalie Badji, 24
Sadio Gnima Badji, 26
Saly Badji
Sarata Badji
Seydou Badji, 23
Seynabou Badji, 19
Souleymane Badji, 43
Souleymane Demba Badji
Souleymane Ngandé Badji
Rokhaya Badji, 25
Mouskoye Badji ep. Tendeng, 43
Rosalie Badjinka, 45
Yoba Badjo, 43
Abdoulaye Baldé, 12
Djadia Baldé, 11
Lamine Baldé
Léonie Baldé
Maodo Baldé, 60
Maodo Francis Baldé, 50
Ousmane Baldé, 22
Cécile Jeanne Bampoky, 47
Cyprien Bampoky, 32
Darry Devy Roland Bampoky, 7
Privé Néma
Félicité Nathalie Bampoky, 18
Gilbert Bampoky
Hyacinthe Bampoky, 23
Joséphine Archangéla Bampoky, 10
Michael Karim Bampoky, 19
Nestor Bampoky, 26

Joseph Marie Bangoura, 21
Auguste Baraye
Augustine Baraye
Antoinette Barboza, 24
Mamadou Lamine Barr, 28
Amy Barro
Zacharia “Zacco” Barro, 22
Bineta Barry
Mamadou Lamine Barry
Samba Diouma Barry, 19
Fernando Basse
Jeanne Basse, 37
Linda Basse, 44
Joseph Pierre Basse Bassène, 33
Aïda Rose Marie Bassène, 12
Albert Bassène, 0
Albert Bassène, 27
Célestine Combe Bassène, 27
Diariba Ousmane Bassène, 4
Emmanuel Bassène, 67
Eric Bassène
Eugène Bassène, 34
Fatou Bassène, 58
Fatou Ndiaye Bassène, 45
Gloria Bassène, 32
Hélène Bassène, 58
Ineste Marie Thérèse Bassène, 15
Isidore Bassène, 21
Jean Bassène, 31
Jean-Bernard Sighgoé Bassène, 21
Joséphine Bassène, 67
LéAndré Bassène
Louis Bassène
Marie Bassène
Marie Boutouaye Bassène
Marie Nicole Bassène, 34
Marième Bassène, 34
Moïse Bassène
Momar Diakaw Bassène, 50
Ousmane Bassène, 3
Philippe Bassène, 32
René Bassène, 36
Richard Bassène, 23
Rodolphe Ambroise Bassène
Théodore Fabien Bassène, 18
Valentin D. Bassène, 30
Yvonne Bassène, 40
Dominique Bassène Coly, 27
Papa Malal Bassène (Dembélé?)
Anna Tyana Bassène Diatta, 21
Joséphine Bassène-Gomez
Adama Bathia, 14
Robert Bathia, 35
Awa Bayo, 31
Khadiatou Bayo
Maïmouna Bayo, 21
Sidy Bayo, 16
Guérande Benjamin, 21
Ndèye Astou Bèye, 25
Angélique Biagui
Basile Biagui, 33
Henri Jacques Biagui, 21
Marcel Biagui
Monique Dominique Biagui, 26
Maria Calmento Biagui, 42
Raphaël Biagui, 20
Siméon Biagui
Mohamadou Lamine Fodé Biaye, 28
Adrien Marthe Bindia
Madiagne Bitèye, 4
Mafall Bitèye, 30
Laurence Blumberger
-- --Bodian, 20
Ata Bodian, 57
Bakary Bodian
Bassirou Bodian, 23
Boubacar Bodian, 28
Ciré Bodian
Malang Bodian
Mamadou Tidiane Bodian, 26
Mariama Michelle Bodian, 6
Moustapha Bodian, 50
Ousmane Bodian
Sanoune Bodian, 22

Sira Bodian, 20
Souleymane Bodian, 45
Astou Aïssatou Bodiang, 42
Félix Boissy, 48
Marie Noëlle Boissy, 20
Elisabeth Birty Boissy-Meyzie, 28
Edmond Nicolas Bora, 31
Penda Boubane
Tim Boury
Moussa Boye, 34
Thierno Mame F. Boye
Francisco Mané Ca, 25
Christophe Cabo
Philipine Moussou Cabou, 17
Mories Maurice Cafone
Aïda Camara
Aïssatou Camara
Amady Camara
Antoine Camara, 25
Cheikh Ibrahima Camara, 8
Fanta Ciré Camara
Fatou Camara, 15
Fatou Camara, 39
Ibrahima Camara, 13
Mamadou Camara
Mamadou Lamine Camara, 14
Mamadou Sy Camara
Mamady Camara
Momath Camara, 35
Sylla Bourama Camara, 21
Yaya Camara, 19
Yaya Camara, 23
Virgine Canalou Camalo, 36
Honoré C. Carvalho
Pierre Marie Carvalho, 25
Rose Anne Carvalho, 26
Barbara S. Chudel
Abdou Ciss
Aïcha Cissé
Bakary Cissé, 21
Boubacar Cissé, 23
Diama Cissé, 24
El Hadji Cissé
Fatou Cissé, 6
Goundo Fatoumata Bintou Cissé, 39
Ibrahima Cissé, 2
Ibrahima Cissé, 22
Insa Cissé, 25
Kéba Elhadji Cissé
Maïmouna Maï Cissé, 39
Mamadou Pape Cissé, 26
Marième Cissé
Ndèye Fatou Cissé, 27
Oumar Boune Cissé, 52
Pape (Mamadou) Cissé
Rama Cissé
Yacine Cissé, 23
Mamadou Lamine Cissokho, 18
Khadidiatou Cissokho, 41
Aïssatou Combé Coly, 46
Alassane Coly, 27
Anastasie Rolande Coly, 7
Awa Coly, 47
Barthelémy Coly, 23
Daniel Coly, 13
Djibril Coly, 11
Emmanuel Alassane Daky Coly, 7
Ernestine Marie Coly, 29
Fallone Coly, 9
Fallou Coly, 7
Fatou A. Coly
Fatou Bintou Coly
Georgette Coly
Ibrahima Joseph Coly, 27
Idrissa Coly
Jean Daniel Coly, 12
Joséphine Coly, 41
Lamine Coly, 10
Lucrèce Elisabeth Coly, 3
Malick Coly, 5
Mariam Coly
Marie Thérèse Laurence Coly, 11
Moussa Coly
Nanou Coly

Nazaire Coly, 34
Oumar Coly, 29
Ousmane Coly, 28
Pape O. Coly
Philippe dit Laurent Coly, 21
Pierre-Marie Coly, 20
Pierre-Marie Coly, 40
Romélie Solange Coly, 23
Sahibatou Silibatou Coly, 46
Samuel Coly, 22
Thérèse Coly
Victor Emmanuel Coly, 32
Younouss Coly, 28
Mamadou Lamine Combo, 26
Bernard Compala
Jean Compalla
Moustapha Conta
Djibril Conté
Antonio Corréa
Alain L. Corréa
Bakary Corréa
Insa Corréa
Mariama Coulibaly, 41
Penda Coulibaly, 50
Marème Couta, 1
Médoune Couta, 2
Alioune Badara Dabo, 23
Djibril Dabo
Fatou Dabo, 5
Malamine Dabo, 0
Malang Dabo, 6
Mamadou Khalifa Dabo, 21
Ramatoulaye Dabo, 44
Yancouba Dabo, 28
Maïmouna Daby
Louise I. Dacogna
Louis Dacosta, 50
Youssouph Daffe, 24
Binta Daffe Yaffa, 22
Ansoumana Dahaba, 31
Faye Dahaba, 34
Sobé Dahaba, 32
El Hadji Danfa
Ibrahima Danfa, 27
Seyni Ndèye Marie Danfakha, 1
Jeanne d'Arc Dasylva, 17
José Dasylva, 47
Khady Dasylva
Clara Gomez Dasilva, 23
Jean G. Decky
Papa Malal Massal Dembélé, 22
Coumba Dème
Djiby Dème
Mamadou Dème, 22
Racky Dème, 13
Boubacar Derwis Davy, 20
Marie Charlotte Désiré
Mamadou Dia
Abdou Dia, 19
Amadou Dia, 25
Fatou Dia, 39
Mohamed Illo Dia, 25
Ousmane Diabang, 17
Ousmane Diaboula, 15
Maïmouna Diaby
Mandack Kandock Diack, 52
Mariama Yama Diack Dia
Jean Diadhiou
Elisabeth Diadhiou
Idrissa Diadhiou / Diédhiou, 23
Adrien Diadia, 26
Grand Yoff Dakar
Fatou Bintou Sébastiene, 39
Momo Diafouné
Abdou Aziz Diagne
Aïcha Liliane Diagne, 13
Ansoumana Diagne, 1
Awa Diagne
Chérif Kéma Diagne, 3
Coumba Mame Coumba Diagne, 1
Fary Sène Diagne, 53
Fatou Diop Diagne, 12
Galaye Diagne, 4
Larissa Dolorès M. K. Diagne, 22

Mactar Diagne, 16
Mamadou Diagne, 18
Mamadou dit Mandiaye Diagne, 9
Mamadou Kébé Diagne, 10
Ndèye Diagne, 37
Ramatoulaye Diagne, 20
Sata Guèye Diagne, 1
Youssou Diagne
Cheikh Tidiane Diaité, 16
Mouhamadou Diakhaté, 22
Amadou Diakhaté
Ibrahima Ba Diakhaté
Makhtar Diakhaté, 27
Mamadou Diakhaté, 32
Aïssatou Diakité
Amadou Diakité
Mariama Diakité
Abdou Diallo
Abdoulaye Diallo, 25
Abdoulaye G. Diallo, 30
Abdoulaye Coco Diallo, 15
Abdourahmane Diallo, 43
Ahmadou T. Diallo, 14
Alassane Diallo, 20
Aliou Diallo, 7
Aliou Diallo, 15
Amadou Diallo, 45
Amadou T. Diallo, 24
Aminata Diallo, 1
Aminétou Diallo, 25
Assane Diallo, 18
Awa Diallo, 37
Awa Diallo, 43
Boubacar Diallo
Cheikh Diallo
Cheikh Ahmed Tidiane Diallo, 15
Codé Diallo
Mamadou Amadou Diallo, 24
Famara Diallo, 0
Ibrahima Niass Diallo
Ismaïla Diallo, 26
Mamadou Alpha Diallo
Mamadou D. Diallo
Mame Diarro Diallo, -1
Mamy Ngoné Diallo, 27
Ousseynatou Ousseynou Diallo, 23
Sadia Faty Samaya Diallo, 24
Safiétou Diallo
Saliou Salif Diallo, 8
Samba Diallo
Souleymane Diallo, 13
Thierno Diallo, 12
Thierno Malam Diallo, 19
Thierno Sabally Diallo, 13
Tidiane Hamadou Diallo, 15
Vieux Diallo
Enfant Diam
Filidié Diambang, 35
Diamilatou Diamé, 13
Moussa Allé Diamé, 40
Fatou Bintou Diandy, 56
Adama Diandy
Aliou Diané, 7
Alioune Badara Diané
Massata Diané
Ousmane Diané
Vieux Diané, 22
Souleymane Diangkol
Ousmane Diao, 24
Fanta Diarra
Mame Fatou Yacine Diarra, 4
Issa Diarra
Moussa Diarra
Fatou Diassy, 16
Abdou Diatta, 21
Abdou Patrick Diatta, 0
Adiambon Diatta, 8
Adrien Diatta, 22
Agnès Diatta, 31
Alain Sidy Diatta
Alassane Diatta, 20
Albert Diatta, 28
Albert Sibindia Diatta, 35
Almamy Diatta, 17

Alphonse Diatta, 22
Amadou Dia Diatta, 25
Ambroise Diatta, 48
Ampa Abadia Diatta, 25
Amy Diatta
Anna Essaloé Diatta, 21
Anne Marie Hélène Diatta, 20
Arlette Signou Alexignou Diatta, 22
Arnaud Armand Diatta, 13
Arona Diatta
Astou Ndel Diatta, 26
Atabodian Diatta, 50
Awa dite Fanseyni Diatta, 36
Baba Diatta, 27
Babacar Diatta
Bacary Diatta, 32
Badou Baba Diatta
Barthelémy M.A. "Batou" Diatta, 0
Bernadette Diatta, 25
Bigué Diatta
Bineta Diatta, 12
Binta Diatta, 52
Bintou Diatta, 36
Bouly Diatta, 48
Caroline Diatta, 7
Cécile Faty Diatta
Cheikh Abdoul Ahad Diatta, 35
Christine Gnima Diatta, 9
Cyriaque Diatta, 22
Delphine Diatta, 31
Diarra Ndiaye dite "Nini" Diatta, 0
Didier Diatta, 20
Diong César Diatta, 22
Elisabeth Alice Diatta, 28
Ernest Diatta, 19
Famara D. Diatta, 4
Fatou Bintou Diatta, 21
Fatoumata Diatta, 2
Fatoumata Diatta, 2
Gabriel Diatta
Gilbert Julbert Jules Arnaud Diatta, 5
Hélène Diatta, 42
Henriette Diatta, 22
Hilaire Tapou Diatta, 33
Hocalackatte Diatta
Ibrahima Diatta, 27
Isidore Diatta, 24
Issa Diatta, 36
Jacques Diatta, 51
Jean Marie Diatta, 39
Jean Pierre Diatta, 19
Jean Pierre Diatta, 43
Jeanne d'Arc Diatta
Joachim Lammuntho Diatta, 10
Joanna (Jiwana / Djiwara) Aline Boka (Booka) Diatta, 22
Joseph Diatta, 44
Joseph Outy Diatta, 10
Lamine Diatta
Léon Pascal Diatta
Léontine Diatta, 29
Louis Diatta, 23
Madeleine Diatta
Maïmouna Diatta, 11
Maïmouna Diatta, 65
Malick Diatta
Mama Diatta
Mamadou Diatta, 24
Mamadou Lamine Diatta, 20
Mamadou Lamine Diatta, 22
Mamina Diatta, 24
Maria Diatta
Mariane Alindiosso Diatta, 16
Marie Diatta, 37
Marie Diatta, 42
Marie Adiambone Diatta, 8
Marie Adiambone Diatta, 25
Marie Augustine Diatta, 48
Marie Augustine Diatta, 48
Marie Bernadette Diatta, 25
Marie Diembéring Diatta, 36
Marie Kélokéléo Diatta, 30
Marie L. Diatta
Marie Rose Diatta

Martha Gnarayokho Diatta, 40
Marthe Diatta, 50
Martine Diatta, 15
Mathieu Diatta, 19
Maxime Diatta
Mélanie Diatta, 30
Michel Diatta, 41
Moussa Diatta, 20
Moustapha Diatta, 15
Moustapha Diatta, 30
Moustapha D. Diatta, 24
Moustapha Racine Diatta, 23
Natacha Diatta, 2
Ndèye Banna Diatta, 56
Odette Diatta
Oletta Diatta
Ousmane Diatta, 43
Papa Sidy Diatta, 21
Paul Diatta, 35
Paul Eugène Diatta
Prosper Yves Diatta, 17
Rachel Yombé Diatta, 0
Ramatoulaye Djarème Diatta, 24
Richard Diatta, 20
Rosalie Diatta, 22
Rosine Diatta
Sadio S. Diatta
Safiétou Diatta, 17
Samaye Jean Pierre Diatta, 51
Samba Diatta, 22
Sarata Diatta, 56
Serge Ekoraye Diatta, 22
Seynabou Diatta
Souleymane Diatta
Stella Rokhaya Diatta, 30
Suzanne Diatta, 11
Tapha D. Diatta
Thérèse Diatta, 16
Thérèse S. Diatta, 32
Thialy Boye Abdoul Ahat Diatta
Thomas Diatta, 0
Victoire Diatta, 20
Victor Diatta, 31
Vieux Diatta, 42
Vieux Babacar Diatta, 14
Vincent Assamba Diatta, 20
Wada Diatta
Wouye Diatta, 43
Yaya Diatta, 22
Yaya Diatta, 28
Yaya Diatta, 34
Yaya Diatta, 36
Yaya Diatta, 39
Yaye Ndeye Binta J. Diatta, 23
Yolande F. Diatta, 22
Youssouph Diatta, 21
Zacary Diatta
Ndack Katy Diattara, 40
Abdoulaye Amadou Diaw
Abdoulaye Marie Diaw, 28
Abdoulaye Pa Diaw, 29
Lissa Diaw
Mansour Chimère Diaw, 27
Marième Diaw
Moussa Diaw
Ousmane Diaw, 13
Ousmane Aliou Diaw
Pape Mamadou Diaw, 21
Aïssatou Diawara, 12
El Hadj Dioncounda Diawara, 19
El Hadj Mamadou Diop Diawara, 33
Abdoulaye Diédhiou, 34
Almamy Diédhiou
Amoul Diédhiou
Amy Diédhiou, fille de Amidou, 6
André Diédhiou, 28
André Diédhiou
André Jean Diédhiou, 20
Angélique Diédhiou, 46
Angouagou Diédhiou, 32
Ansoumana Diédhiou
Awa Diédhiou, 20
Bakary Diédhiou
Benoît Diédhiou, 4

Binta Diédhiou
Charlotte Diédhiou
Diariétou Diédhiou, 24
Diénéba Diédhiou, 38
Diminga Sambou Diédhiou, 44
Eléonore Marie Diédhiou, 34
Elisabeth Diédhiou
Emmanuel M. Diédhiou, 34
Enfant Diédhiou, 0
Etienne Diédhiou, 34
Fanta Diédhiou, 54
Fanta Diédhiou
Fatou Diédhiou, 10
Fatou Diédhiou, 63
Fatou Kiné Diédhiou, 18
Georges Diédhiou, 28
Georges Pompidou Diédhiou, 22
Ibou Diédhiou, 8
Ibrahima Diédhiou, 30
Ibrahima Diédhiou, 44
Ibrahima Solo Diédhiou, 40
Idrissa Diédhiou
Julien Diédhiou, 27
Juliette Diédhiou, 24
Khady Diédhiou, 5
Lamine Diédhiou, 24
Landing Diédhiou, 23
Landing Diédhiou, 24
Landing Papa Landing Diédhiou, 25
Léonard Diédhiou, 23
Makha Diédhiou
Malamine Diédhiou, 39
Malang Diédhiou, 25
Malang Diédhiou, 47
Mamadou Lamine Diédhiou, 38
Marcel Diédhiou, 15
Mariama Diédhiou
Marie Diédhiou, 40
Marie Eléonore Diédhiou
Mata Martha Diédhiou, 42
Mouhamed S. Diédhiou
Mouminatou Diédhiou, 3
Mous(sou)kéba Diédhiou
Nafissatou Diédhiou
Néné Diédhiou, 22
Nestor Diédhiou
Oulimata Diédhiou, 21
Oumar Diédhiou
Pape G. Diédhiou
Raby Diédhiou
Roger Kéba (Kaba) Diédhiou, 44
Rose Diédhiou, 22
Safiétou Diédhiou
Salif Diédhiou, 20
Séni Diédhiou, 81
Seynabou Ndèye Diédhiou, 45
Seyni Landing Demba Diédhiou, 10
Youssoupha Diédhiou, 22
Adèle Diémé, 27
Adja Diémé
Aïssatou Diémé, 14
Albertine Henriette Diémé, 2
Alpha Alhoussaïny Diémé, 14
Baye F. H. Diémé
Bintou Binette Diémé, 26
Cheikh Omar Diémé, 14
Domitilla Fotienne Diémé, 25
Dona Amalaka Diémé, 1
Fatou B. Diémé
Henri Diémé
Ibrahima Diémé, 9
Ibrahima Diémé, 63
Ibrahima Pape Diémé, 25
Isaïe Fernand Diémé, 40
Khady Diémé, 40
Kory Diémé
Léoure Diémé, 25
Mamadou Diémé, 13
Mamadou Basséna Diémé, 6
Mariama Diémé, 14
Marie Diémé, 22
Marie Hélène Diémé, 20
Matar Makhtar Diémé, 57
Moussa Diémé

Ndèye Khadidiatou Diémé
Pierrette Diémé, 58 ans, fille de Henry et Térèse Sambou, Ménagère
Prince Bacary Diémé, 9
Racky Diémé
Ramatoulaye Diémé, 35
Serge Ingri Diémé, 21
Serigne Mbaye Ababacar Diémé, 15
Yaya Diémé, 43
Paul Diène, 30
Abdoulaye A. M. Diène
Abibou Habib Dieng
Aïssatou Aïchatou Dieng
Aliou Dieng, 44
Assane Dieng, 21
Assane, enfant Dieng
Cheikh Dieng, 25
El Hadji Dieng, 24
Fatou Assane,enfant Dieng
Ibrahima Dieng, 23
Samba Dieng, 47
Sata Dieng
Véronique Dieng, 26
Mariama Dieudame, 40
El Hadji Dièye, 34
Fatou Dièye, 8
Fatou Dièye, 9
Galaye Dièye, 40
Ibrahima Dièye
Magui Dièye
Abdoulaye Diol, 25
Léna Diompy
Lina Diompy
Alain Dione
Ousmane Dione
Baïlo Dione
Thiome Dione Diom, 67
El Hadji Dione Ndione, 40
Lamine Diongue, 29
Elisabeth Dionou, 20
Mamadou Diop, 45
Abdou Diop, 3
Abdoukhadre Diop, 5
Abdoulaye Diop, 49
Abibatou Diop, 47
Alassane Demba Diop, 43
Aminata Diop, 36
Aminata Diop, 50
Amsata Diop, 7
Awa Diop, 33
Awa Thiam Diop, 22
Binta Diop, 5
Bintou (Bineta) Diop, 19
Cheikh Bamba Diop, 4
Core Cou Coco Diop, 47
Coumba Diop
Demba Diop, 41
Demba Diop, 44
El Hadji Mamadou Diop
Fatou Cissé Diop, 40
Ibrahima Diop, 43
Insa Diop, 22
Kabir Karim Diop, 2
Khady Khadidiatou Diop, 52
Libasse Diop
Malick Gorguy Diop, 25
Mamadou Diop, 6
Mame Bamba Diop
Marième Diop
Mohamed El Bachir Diop, 28
Mountakha Diop
Moussa Diop, 11
Moustapha Diop, 32
Ndèye Maïmouna Diop
Omar Diop
Pape Sémou Diop, 14
Papis Abdou Diop, 5
Salimata Diop, 4
Saly Diop
Souleymane Diop, 7
Souleymane Diop, 26
Souleymane Diop
Tafsir Amar Momar Diop, 36

Thierno Diop
Ida Idrissa Diop Diouf, 23
Abdou Karim Diouf
Amath Diouf
Emma Diouf, 20
Emmanuel Diouf
Ibrahima Diouf, 40
Khady Diouf, 30
Mayacine Diouf, 47
Ndéné Diouf
Ndèye Sira Diouf, 34
Ousmane Diouf, 38
Marie Diouf Dieng, 39
Léontine Dioukane (Ndioukane), 36
Aïssatou Dioum
Amadou Alé Aly Dioum, 35
Baye Amadou Dioum
Agdam Abdou Djako Djato
Daba Rokhaya Djiba, 10
Daouda Djiba, 24
Hélène Djiba, 22
Hélène Djiba
Isabelle Djiba, 14
Isabelle Victorine Djiba, 13
Karim Djiba, 22
Malamine Lamine Djiba, 24
Mamadou Djiba
Mamadou Lamine Djiba, 32
Matar Makhtar Djiba, 12
Michel Rigobert Djiba, 10
Oumy Djiba, 28
Sébastien Djiba, 48
Henry Djiboune, 19
Albert Djicoune, 4
Hyacinthe Djicoune, 24
Youssouph Djignaly, baby
André Casimir Dieudonné, 18
Paoulino Dju
Mamadou Doumbia, 27
Mariame Doumbouya, 20
Ousseynou Doumbouya
Lola Lala I. Ralla Doumbouya, 16
Jean Pierre Dousset, 37
Adama Dramé, 57
Amadou Dramé, 21
Amissatou Abissatou Dramé, 23
Ansou Dramé, 27
Fatou Guèye Ndiaye Dramé, 9
Fatou Hodio Dramé
Mohamed Cheikh Dramé, 21
Oudia Dramé
Oumar Dramé, 16
Pape Dramé, 27
Ramatoulaye Dramé, 60
Souleymane Dramé
Vieux Dramé
Yaye Fatou Dramé, 8
Sounkarou Dramé Dawo, 19
Aliou Dramé Diané, 4
Dorothée S. B. Duchateau
Paqui Pecky Pegguy M. L. Edet
Stephan Essien, 32
Fatoumata Fabouré, 50
Alpha Fadiaba Famialo, 22
Ahmet Fall
Amadou Makhata Fall, 21
Awa Fall, 41
Babacar Fall
Bakary Fall, 7
Cheikh Kheuss Fall, 27
Idrissa Fall, 76
Mamadou Fall
Marame Fall
Mbagnick Fall
Mouhamed Moustapha Fall, 0
Ndongo Fall
Ngadis Fall
Pape Fall, 22
Pape Cheikh Fall
Tamsir Fall, 9
Tida Fall, 43
Cécile Faty, 15
Lamine Faty
Mamady Faty, 16

Aliou Faye, 26
Aliou Badara Faye, 30
Baye Mayacine Faye, 47
Diarno Diouma Faye, 53
Faly Diop Faye
Ismaïla Faye, 26
Kadiatou Faye, 33
Maguette Faye, 14
Malick Faye, 10
Mame Mbissine Sène Faye, 5
Mariane Faye
Mariétou Agnès Faye, 24
Mayacine Faye, 27
Mbath Samba Faye, 9
Oumar Faye, 12
Oumar Diop Faye
Ousmane Faye, 46
Papa Waly Faye, 28
Pape Makha Faye, 20
Saliou Faye
Waly Faye
Nyanga Gnagna Faye Sène, 13
Anastasia Fernandez, 23
Jean André Simon Fernandez, 19
Marie Fernandez
Céleste Figueira, 26
Mariama Fodera Fadera, 20
Kinté Kiné Lopez Dafonseca, 50
David Daniel Forbé
Léontine Gandoul, 30
Abdoulaye dit Mafaly Gassama, 25
Balla Moussa Gassama, 14
Coumady Gassama
Zéneide D. Gassama
Marie Gaye, 38
Vieux Gaye
Ndiawar Gaye
Cheikh T. Y. Gning (Gningue)
Ibrahima Gningue, 24
Souleymane Goly, 16
Alain Gomis, 35
Albert Gomis
Alfred Gomis, 25
Alissa Gomis
Anna B. K. Gomis, 30
Aristide Gomis
Baptiste Bas Pintou Gomis, 58
Bas Pintou Gomis, 68
Catelle Gomis
Denis Gomis
Désirée R. Gomis, 29
Elie Gomis, 25 ans
Elisa Bayo epouse Sylva Gomis, 54
Euclidis Euclide Gomis, 22
Florence Florentine Gomis, 36
François Gomis, 25
Geoffroy Gomis, 27
Germaine Gomis, 20
Gilbert Gomis, 27
Henry Gomis
Jean Bosco Gomis, 15
Jean Paul L. Gomis, 14
Jean R. Gomis
Jean-Louis Gomis
Jean-Marc Gomis, 19
Jean-Paul Gomis, 13
Joachim Gomis, 14
Khady Gomis
Libasse Gomis
Mangueka Gomis, 45
Marie Hélène Gomis, 20
Maurice Gomis, 12
Maxi(me) K. Gomis, 17
Oupa Nana Gomis, 40
Paul Gomis, 49
Pauline Gomis
Philomène Gomis, 23
Rachel Gomis, 3
Roumon Romane Ronco Gomis, 25
Sylvain Gomis, 40
Sylvain Valéra Gomis, 23
Victor Gomis, 28
Victor Gomis, 57
Virginie Caréda Gomis, 22

Yacine Bosco Gomis, 15
Carlos Gomis Gomez, 18
Maria Kougna Gomis Gomez, 31
Marie Gomis Gomez, 20
Antoine G. Gomis Gomez, 12
Francisco dit Thicou Gomez, 43
François Gomis Gomez, 45
Abdou Goudiaby
Abdourahmane Goudiaby, 18
Aminata Goudiaby
Bakary Goudiaby, 18
Bakary Goudiaby
Bakary Vieux Goudiaby, 33
Delphine Goudiaby, 16
Doudou Goudiaby, 50
Ernest Bedouin Goudiaby, 16
Fatou Lala Goudiaby, 24
Gnimading Goudiaby, 17
Ibrahima Goudiaby
Isabelle Michèle Goudiaby, 23
Issa Goudiaby, 23
Mame Diarra Goudiaby, 1
Mouminatou Goudiaby, 43
Moustapha Goudiaby, 21
Papa Ousmane Goudiaby, 26
Penda Goudiaby
Ramatoulaye Goudiaby, 27
Sophie Prisca Goudiaby, 24
Tombon Goudiaby
Corine Guedj
El Hadj Ma Ngoné Guèye, 7
Abdoulaye Guèye, 24
Adama Guèye, 34
Amadou Guèye
Assane Guèye
Bineta Guèye, 20
Coumba Diop Guèye, 0
Coura Guèye, 9
El Hadj. Mor Khoudia Guèye, 21
Ibrahima Guèye, 26
Idrissa Guèye
Magatte Magui Guèye, 30
Maïmouna Guèye
Malick Guèye
Mamadou Guèye
Mamadou Ba Guèye
Marie Guèye
Ndèye Aïssatou Guèye
Omar Guèye, 22
Papa Guèye, 8
Pape Malick Guèye, 27
Samba Guèye
Abdou Abou Hassane
Abdou Karim Houndjo, 27
Christine G. Humeau Suzanne
Nennura Janice Mendes Indami, 18
Frenil Funil Indy
Jean Louis Izard, 1
Gérard Jacky, 40
Justin Jacky, 41
Adeline Addine Diatta Sané, 22
Piennely Jeannette
Jorge Diaz de Tudanca Jimenez
Marguerita Salvador Jimenez
Maria Diaz de Tucanda Jimenez
Naly R. Nelly P. Joseline
Aristide Bernard Charles Jules, 62
Amadou Ka, 26
Elhadji Mamadou Ka
Elimane Mamadou Ka
Fatoumata Ka, 37
Hamady Amadou Ka
Moussa Ka, 30
Ndiaga Ka, 41
Alioune Kabo
Barthélemy Kabo
Charlemagne Kabo
Jean R. Kabo
Antoine Kabo, 23
Antoinette Kabo
Marie Kabo, 50
Zène R. Kabo
Alice Augustine Kabou, 14
Luc Pascal Kabou, 14

Adèle Konady Kabou, 23
Antoine Kabou
Karla Kabou
Marie Thérèse Kabou, 33
Eloge Milliam Viviane Kagnaly, 22
Malaine Madeleine Kama, 8
Momar Guèye Papa Kama, 5
Virginie Kamala, 32
Christian V. Kamara
Ibrahima Kamara, 21
Momath Kamara
Yaya Kambaye, 25
Lamine Kandé
Robert Kandetty, 42
Adama Kandji, 29
Diaba Kane
Fatoumata Kane, 22
Fatoumata Kane, 63
Iba Kane, 49
Ibra Kane, 55
Kaya Demba Kane, 27
Mame Diarra Kane
Bernabé Joseph Kanfom Rangom
Moïse Kanfon, 23
Célestin Kanfoudy, 32
Ibou Kanté
Saturnin Kantoussan
Albert Kassoka, 32
Chompy Sandrine Kassoka, 20
Isabelle Kassoka, 29
Rosalie Kassoka, 15
Alfred Charles David Kayounga, 6
Suzanne Sylvie Kayounga, 15
Véronique Béatrice Kayounga, 14
Abdou K. Kébé
Adama Kébé, 8
Alioune Kébé, 35
Awa Kébé, 8
Farakh Sarakh Kébé, 6
Mamadou Kébé, 3
Moussa Kébé, 4
Bassirou Keita
Cheikh Sadio Keita
Cheikhna Keita, 26
Djiby Keita, 24
Ibrahima Keita, 23
Gaston Keita
Achille Kény, 24
Sénébou Khoulé, 22
Léontine Ko Nkoum, 28
Madeleine Coumba Koïta, 28
Malang Koma, 36
Mamadou L. Kombo
Ousseynou Konaté, 17
Seydou Koné
Fatoumata Konté, 49
Alioune Badara Konté, 22
El Hadj Ibrahima Konté Kanté
Abdoulaye Korta, 48 ans
Penda Korta
Samba Kouaté, 37
Daouda Koundoul, 42
Oury Kouroum
Samba Kouyaté
Grace Ansilémoloute Lambal, 1
Lysiane Manboutine Lambal, 7
Marie Lambal, 38
Yves Marc Amaga Lambal, 5
Blandine M. T. (M.E.) Lambal
Ndeye Marième Faye Lamotte, 23
Domizio Lepoci Lipoci
Ndambert Lèye
Awa Lodiane Badiane, 32
Joseph Siacrite Lomène, 10
Béatrice Lopes, 23
Beatriz Lopez, 36
Carlinoiu Lopez, 10
Emmanuel Roger Lopy, 66
Fernand Baraye Bamaye Lopy, 20
Richard Lopy, 28
Néné Lopy Lopez, 37
Néné Lopy Lopez, 39
Fatou Ly
Ibrahima Ly, 27

Assomption Malack, 45
Abraham Malomar, 1
Edmond Malou, 27
Martin Malou, 27
Pascal Luc Pascal Malou Kabou, 15
Oulda Gilda Malou, 38
Angélique Mancabou Macalou
Cyprien Mancor, 33
Serge Hubert M. Mandiamy, 28
Maguette Mandiang, 57
Alonze H. M. Mandione Mandiami
Adama Mané, 43
Aïssatou Mané
Aladji Mané, 26
Aliou Mané
Aminata Mané, 57
Ansoumana Mané, 13
Astou Mané
Awa Mané, 38
Binetou Mané, 18
Christophe Mané, 25
Daouda Mané, 37
Dionsignou Mané, 25
Djiwoya Thérèse Mané, 24
El Hadji M. Mané, 20
El Hadji Mamadou Mané, 16
Fally Mané
Fatou Mané, 11
Fatou Mané, 58
Fatoumata Mané, 9
Jean Paul Mané, 12
Kémo Mané, 24
Khady Mané, 36
Lamine Malamine Mané, 11
Léna Léa Mané, 46
Ma-Awa Mané
Malang Mané, 25
Ousmane Mané
Philomène Tacky Mané Mané, 27
Raimundo Mané, 51
Rose Preira Mané
Seynabou Mané, 2
Sokhna Mané, 18
Souleymane Mané, 45
Victorine Mané
Mamadou Mané Diafoune, 3
Nfally Mané Diagne, 48
Adamasse Manga, 57
Alain Manga
Alouga Mbinki Manga, 4
Alphonse Manga
Amy Manga, 36
Anna Anta Lidjie Manga, 20
Brigide Prigibbé Manga, 19
Bruno Manga, 28
Camille Manga, 8
Césard Cédar Nango Manga, 12
Charles Manga, 16
Charles Manga, 56
Damas Manga
Ermande Hermann Manga, 25
Ernest Benoit Manga, 16
Eudoxie Manga, 28
Fatou Manga, 18
Gabriel Benjamin Manga, 26
Gilbert Abdou Lahat Manga
Gilbert Edougboungué Manga, 62
Hernès Hermann Germain Manga, 5
Hervé Kambing Manga, 23
Idrissa Manga, 27
Jean-Louis Manga
Jean-Pierre Manga, 0
Joséphine Manga
Judith Manga, 33
Kempès Mario Kempes Manga, 1
Kombé Manga, 6
Madeleine Manga, 4
Marcel Manga, 24
Marcel S. Nango Manga, 3
Marcelino Manga, 5
Maria Manga, 41
Marie Manga, 8
Marie-Jeanne Manga, 4
Micheline Manga, 0

Nély Valérie Manga, 1
Odokite Manga, 30
Ousmane Manga, 0
Pascal Manga, 34
Pisco Manga, 3
Raoul Manga, 18
René André Louis Manga, 16
Rosalie Siabatou Manga, 1
Rosine Manga, 36
Saliou Manga, 24
Sarata Manga, 1
Serge I. Manga, 26
Baïdy Makier Mankane, 18
Abdou Mansaly, 21
Ariel Giscard Mantounck, 11
Charles Pascal Mantounck, 15
Joseph Mantounck, 17
Olive Suzanne Mantounck, 22
Guy Achille Ousmane Mansaly, 10
Fayrouze Thérèse Maouad, 39
Youssou Maréna, 23
Youssoupha Maréna, 32
Jean P. G. Marie
Dominique Ibrahima Marone, 18
Atagnassio Martin
Marcel K. M. Martin
Joséphine Désirée N. Maty, 31
Alpha Oumar Mballo, 22
Mamadou Bakary Mballo, 44
Adji Bigué Mbaye, 24
Boubacar Mbaye, 23
Chérif Younouss Mbaye, 43
Diétou Mbaye, 39
Ibrahima Mbaye, 40
Idrissa Mbaye, 18
Insa Mbaye, 30
Mariama Mbaye
Ndèye Kenbougoul Mbaye, 24
Pape Alioune Mohamed Mbaye, 27
Serigne Mbaye
Mbaye Mbengue
Papa Khaly Aly Mbodj, 18
Mame G. Mbodj
Mamadou Mbodji
Djiba Djiby Mboup, 48
Everlast Arnaud Mbourou, 26
Julien Médou, 40
Armindo Mendes, 20
Louis Mendes
Nessiaguira D. Mendes
Awa Anna Mendes Mendy, 2
Fernando Mendes Mendy, 44
Adèle Mendy, 26
Alain Dion Marcel Mendy, 27
AlexAndré Mendy, 23
Alphonse Mendy, 44
Antoine Mendy
Armando Mendy, 21
Bonaventure Mendy, 4
Célestin Mendy, 27
Charles Mendy, 38
Daniel Mendy, 12
David Mendy, 12
Denis Mendy, 15
Diéto Mendy, 22
Diminga Mendy, 70
Domingo Mendy
Elisabeth Mendy, 2
Fatima Fatou Mendy, 37
Florine Floris V. Mendy
Gaston Mendy, 9
Geneviève Anorane Mendy, 3
Georgette T. Mendy
Gérard Mendy, 23
Hélène Mendy
Hortense Gomis Mendy, 31
Jean Pierre Mendy, 25
Jean L. E. Mendy
Jeannette Mendy, 32
Laurent Mendy, 20
Laurent Mendy, 47
Lazare Mendy, 13
Lazare Mendy
Léna Mendy, 20

Léon Mendy, 14
Loty Louis Mendy, 23
Madeleine Mendy 39
Mandithiss Mendy
Marcel Mendy
Marie Mendy
Marie Thérèse Mendy, 31
Nelson Mendy, 16
Nganthiou Mendy
Pape Charles Mendy, 15
Pathiala Mendy, 46
Patrice Mendy, 22
Paul Mendy
Sabelle Mendy
Samuel Mendy, 6
Sébastien Mendy, 2
Sera Mendy, 23
Siga Mendy, 29
Thérèse Néné Mendy, 30
Trissia Mendy, 2
Odile Mendy Diédhiou
Monique Mendy Mendes, 32
Jacques Meyzie, 38
Sylvia Sylvaine Mingou
Emmanuel Mingou, 47
Jean Yves Miranda, 27
Joya ou Yoya Joel Monteiro, 28
Nelly Morin, 46 ans
Feyrouze T. Mourid Mouad
Anna Namatane
William Nassalan, 19
Marie Louise Nassalang, 36
Michel Rigobert Natack, 16
Paul Natack, 39
Abdoul Wahab Ndao, 3
Lissa Ndao, 0
Ndèye Fatou Ndao, 12
Oumou Oumy Ndao, 34
Ousmane Ndao, 23
Marie-Emilie Ndao, 0
Marième Mané Ndao, 0
Auguste Ndecky, 62
Bernard Ndecky, 41
Christine Ndecky, 49
Jean Ndecky
Jean Pierre Ndecky
Rosalie A. Ndecky, 34
Cécile Ndecky, 30
Charlotte Ndecky, 51
Fernand Ndecky, 45
Marie Ndecky, 24
Henriette Ndeye, 59
Abdoulaye Ndiaye, 33
Alassane Ndiaye, 19
Aliou Ndiaye, 25
Amadou Ndiaye, 18
Amadou Alhousseynou Ndiaye, 24
Amady Ndiaye, 28
Angèle Maman Ndiaye, 28
Astou Ndiaye
Astou Angèle M. Ndiaye, 20
Awa Ndiaye, 12
Awa Ndiaye, 17
Babacar Ndiaye, 14
Bineta Ciré M. Ndiaye
Binta Ndiaye, 21
Boubacar Ndiaye
Cheikh Ibrahima Ndiaye, 30
Cheikh O. Ndiaye
Coumba Diop Ndiaye, 15
Diaratou Ndiaye, 22
Diarra dite Mimi Ndiaye, 17
Diary Ndiaye, 20
Dieynaba Ndiaye, 19
Dyénaba Ndiaye, 17
El Hadji Ndiaye, 30
Farmata Moussa Ndiaye, 43
Fatou Ndiaye
Fatoumata Moussa Ndiaye, 36
Ibrahima Ndiaye, 25
Ibrahima Ndiaye, 27
Idieu Ndiaye
Idrissa Ndiaye, 26
Maguette Ndiaye, 40

Maïmouna Ndiaye
Mama Ndiaye, 30
Mamadou G. Ndiaye
Mame Sémou Ndiaye, 22
Mamour Ndiaye
Mariama Ndiaye, 17
Mariama Ndiaye, 21
Marie Ndiaye
Masse Ndiaye, 6
Matar Ndiaye, Maître
Modou Ndiaye, 20
Modou Ndiaye, 40
Mohamed Bachir Ndiaye, 17
Momar Ndiaye, Commerçant
Moussa Ndiaye, 18
Moussa Ndiaye, 35
Moustapha Ndiaye, 45
Ndèye Ndiaye, 37
Ndèye Fatou Ndiaye, 22
Ndèye S. C. Ndiaye
Ousseynou Ndiaye, 23
Pape Mahmoud Ndiaye, 15
Pape Matar Ndiaye, 40
Pape Oumar Ndiaye, 15
Rabiatou Ndiaye
Samba Ndiaye
Serge Alioune Ndiaye, 25
Seynabou Ndiaye, 14
Seynabou Ndiaye, 17
Youssou Ndiaye
Ndèye Astou Ndiaye, 25
Mohamed Bachir Ndiaye Diakho, 23
Momar Ndiogou
Bruno Serge Ndione, 35
Rivaldo ou Rivalda Ndioukane, 4
François Ndong
Ndèye Astou Ndong, 24
Mariétou Ndongo
Awa Ndour, 33
Cheikh Oumar Ndour, 13
Jean-Jacques Félix Ndour, 0
Maty Ndour, 29
Ousmane Ndour
Amélie Ndouye, 53
Cheikh Ndoye
Lucien Eric Ndouye, 24
Alfred Ndy, 32
Nicolas Nene
Momar Ngiogou
Marie Charlotte Désirée Ngombé, 31
Adama Ngom
Moussa Ngom, 31
Ousmane Niabaly, fils K Diabaly
Fernando Niaka, 23
Jean Marie Niandio, 18
Ahmed Niang, 25
Amadou Mahamidou C. Niang, 25
Niang, Commerçant
Amadou Kor Niang, 26
Ameth Tidiane Niang, 38
Mame Atou Niang, 46
Ibrahima Niang, 19
Jean Marie Niang
Mamadou Niang, 26
Mbassa Niang, 18
Mbaye Niang
Modou Niang, 55
Sérigne Niang
Thierno Faly Niang, 40
Tidiane Hamad Niang, 37
Mamadou Lamine Niassy, 28
Wahana Samsudin Nias, 23
Oumy Mamy Sossa Niassy, 34
Jean-Marie Niavale
Nicolas Niouky, 48
Rodrigue Arsène Niouky, 24
Fernandou L. Njai
Evelyne Nougha, 26
Euline Nouma
Christophe Ntab, 26
Marie Ntab, 46
Marie Andrée Ntap, 29
Clotaire Elantène Nunez, 53
Joseph François Nunez, 71

Sandrine Satou Nunez, 31
Antoine Nyafouna
Marcellin Jean Nya Fauna, 14
Maxime Nyafouna, 30
Pierre Pape Sougné Nyafouna, 21
Ingheal Inghild Malkanaes, 25
Jean Claude François Nzalé, 24
Grégoire Nzalé, 33
Richard Nzalé, 25
François Nzalé, 24
Justin dit Papis Nzalé, 27
Gersin Fulgence Nzaly, 5
Estelle Nzaly, 10
Zé Oliveira, 25
Joséphine Orange
Alexis Oudiane, 29
Alexis Oudiane, 38
Kadiata Khardiata Dembo Pam, 56
Cynthia Marlène Sintia Pereira, 1
Marie-Nicole Pereira, 35
Dominique Michel Peroni, 20
Bernard Preira
Dominique Preira
Elisabeth Preira, 23
Jean B. Preira
Kamanso Kamapro Preira, 15
Léontine Preira, 50
Marie Preira, 44
Martha Preira, 59
Oupathiam Opa Preira, 19
Rosalie Preira, 67
Rose Preira, 42
Custodia Pereira de Carvalho
Abdoulaye Sabaly, 18
Dieynaba Sabaly, 25
Mariama Sabaly, 35
Alkaly Sadio, 14
Anna Sadio
Aramata Ramatoulaye Sadio, 10
Awa Sadio, 21
Charles Zachariah Sadio, 0
Chérif Younouss Sadio, 16
Dianké Mane Sadio, 51
Fatou Sadio, 44
Fily Philippe Sadio, 17
Lamine Sadio, 18
Léa Anna Sadio, 1
Malang Sadio
Mamadou Lamine Sadio, 22
Rose Sadio
Saly Sadio
Téning Sadio, 2
Yaya Mabo Sadio, 24
Saléhédine Safiedine
Abdoulaye Sagna, 44
Adélaïde Sagna, 25
Adolphe Jules Olivier Sagna, 21
Aïssatou Combé Sagna, 34
Aladji Sagna, 7
Alassane Sagna, Etudiant
Albertine Titina Sagna, 12
Amy Sagna, 11
Anicet Sibédiène Sagna, 24
Ansou Sagna, 12
Antoine K Sagna, 20
Arone Sagna
Astou Maman Sagna, 16
Athanase Michel Sagna
Boubacar Sagna, 30
Casimir Sagna, Soldat 1re classe
Chérif Boubacar Sagna, 26
Clarisse Sagna, 8
Clarisse Sagna, 20
Edouard Sagna, 36
Fatoumata Sagna, 15
Flavien Sagna, 21
François Soubéroune Sagna, 3
Hyacinthe Sagna, 32
Jean-Jacques Sagna, 23
Joachim Lammuntho Sagna, 17
Joséphine Sagna
Khady Elisa Sagna, 20
Louis Isidore Sagna, 26
Luc Assé Dominique Sagna, 8

Maïmouna Sagna, 12
Mariama Sagna, 18
Marie Michel Sagna
Marthe Raïssa Sagna, 21
Massy Sagna, 15
Michel Sagna
Moussa Bassène Mamadou Pape Moussa Sagna, 4
Ndoye Sagna, de Yaya et Mariama Sané
Oumy Ourye Sagna, 33
Pape Oumar Sagna
Pierre Dominique Sagna, 17
Pierrette Hady Sagna, 24
René André Sagna, 25
René Louis Sagna, 11
Stanislas Sagna, 23
Viviane Rose Sagna, 23
Sékou Cheikhou Sagna, 24
Alassane Djigo Sakho, 9
Madina Sakho, 13
Makhan Sakho, 6
Aïssatou Aïssata Mamadou Sall, 6
Alassane Boubacar Sall, 6
Alioune Sall
Fatimata Boubacar Sall, 7
Fatoumata Bintou Adja Sall, 31
Hadji Boubacar Sall, 36
Mamadou Sall
Mamadou Wally Sall, 15
Mbéry Mory Sall, 31
Mory Sall
Oumar Sall, 19
Oumar Idy Sall, 18
Antoine Sam
Pierre Sam
Alhousseynou Samaké, 23
Amy Samb
Assane Samb
François Samb
Mamadou Samb
Ousmane Samb
Joseph AlexAndré Samba, 20
Ahoudy Sambou, 33
Alphonse Sambou, 30
Aminata Sambou
Anna Sambou, 28
Bernard Sambou, 32
Bertrand Sambou, 30
Cheikh Anta Sambou, 14
Diatou Sambou, 16
El Hadji Sambou, 27
Ephrem Sambou, 48
Ferdinand Nicolas Sambou, 38
Henri Boris Sambou, 14
Hortense Sambou, 45
Jean-Marie Sambou, 42
Jeanne Joanna Sambou, 36
Lamine Sambou, 51
Louis Cobie Tobie Sambou, 28
Louis Tobie Sambou, 36
Mama Sambou, 60
Mamadou Eric (L.) Sambou, 28
Moustapha Sambou, 44
Moustapha Sambou
Narcisse Sambou, 14
Odile Sambou
Oumar Sambou, 23
Philippe Sambou, 26
Prudence Sambou, 15
Samory Kabila Arikh Sambou, 10
Sougoute Caroline Sambou, 1
Thérence Sambou, 11
Valère Sambou, 27
Victoire Sambou, 55
Victoria Sambou, 29
Zacharie Sambou, 8
Justin Sambou Diédhiou, 44
Bineta Binette Sambou, 35
Marang Sandé
Moussa Sanding, 26
Abdou Sané, 50
Adama Sané, 24
Aïssatou Sané
Aladji Sané
Aliou Sané, 11

Alioune Sané
Aloyse Sané, 8
Alphousseynou Sané, 22
Astou Siyabatou Sané, 31
Babacar Sané, 22
Bana Sané, 9
Bocar Ba Sané, 25
Boubakar Sané, 35
Bouba Sané, 18
Boubacar Sané, 20
Boubacar Sané, 25
Cheikh Fodé Sané, 20
Diorama Diorama Sané, 26
Emmanuel Sané, 6
Fatou Sané, 12
Fodé Sané, 43
Gnima Sané
Hyacinthe Charles Toussaint Sané,63
Ibrahima Sané, 18
Idrissa Sané
Kaya Demba Sané
Khady Sané
Lamine Sané, 19
Landing Sané, 60
Maïmouna Sané, 43
Mamadou Lamine Sané, 30
Mamadou Sellé Sané, 17
Mariama Sané, 50
Nouha Sané, 15
Ousmane Sané, 15
Ousmane Kader Sané, 13
Seynou Sané
Yaya Sané, 25
Yaye Fatou Sané
Mamadou Sanghare
Bébé Sanka, 27
Moulay Sankaré
Mohamadou Sanokho, 35
Claude Santos, 39
Abdou Khadre Sarr, 27
Anne-Marie Sarr, 46
Arouna Sarr, 42
Awa Sarr, 23
El Hadji Mamadou Sarr, 10
Fatou Sarr, 26
Fatou Sall Sarr, 30
Fatoumata Sarr
Mamadou Sarr, 10
Mariama Sarr, 38
Marie C. Sarr
Mariétou Sarr, 18
Martin Sarr, 28
Mouhamadou Lamine, 6 maanden
Moussa Sarr, 11
Nicolas F. Sarr
Pascal Sarr
Petit Sarr
Tidiane Sarr
Tidiane Salokassine Sarr, 18
Khady Seck, 46
Abdou Abdourahmane Seck, 19
Alioune Sène, 40
Babacar Sène, 26
Boubacar Sène
Djibril Sène, 32
Fary Sène
Ibrahima Sène
Joseph Birame Sène, 28
Lamine Sène
Mame Sène
Mouhamadou Lamine Sène, 30
Mouhamadou Moustapha Sène, 33
Diome Sène Sané
Nouha Sène Sané
Gnima Sène Sané
Ngoné Sène-Izard Sané, 34
Anatole Senghor, 18
Arsène Assingalène Senghor, 21
Catherine Senghor
Cécile Senghor, 29
Constance Senghor, 7
Fatou B. Senghor
Henriette Senghor
Hyacinthe Senghor
Jacqueline Senghor, 26
Jean Djibalène Senghor, 27

Jean Senghor
Jean Laurent (Jeannot) Senghor, 16
Jeanne d'Arc Senghor, 19
Kouloukou Suzanne Senghor, 18
Mignone Senghor
Nathalie Senghor, 14
Rabiline Senghor
Valère Senghor, 33
Abdoulaye Seydi, 27
Amadou Seydi, 30
Cheikh Atab Seydi, 28
Cheikh Mamina Seydi, 28
Ibrahima Seydi, 25
Mamadou Seydi, 40
Mamadou Lamine Seydi, 29
Mbaye Codé Seydi
Pape Salif Seydi, 10
Penda Seydi
Banda Sèye, 10
Gnagna Sèye, 4
Mbaye Sèye, 27
Débit Sèye, 42
Yoro Moulaye Sèye, 48
Adama Sidibé, 55
Fatou Binetou Sidibé Seydi, 20
Baba Simal
Henri Sina, 42
Songoutou Sissokho, 11
Joseph Sleiman, 14
Sékou Cheikh Souleymane Soly, 11
Adama Sonko, 1
Mahomet Sarany Safiétou Diédhiou
Elisabeth Sonko, 37
Maïmouna Sonko, 11
Mamady Sonko, 16
Sabel Sonko, 37
Sabily Sonko, 35
Toumany Sonko, 28
Amy Souané, jonger de 1 maand
Aminata Souaré
Téninké Souaré, 12
Moussa Soumaré, 26
Abdoulaye Binta Sow, 65
Amadou Sow
Arona Sow
Cheikh Ahmed Tidiane Sow, 38
Dié Diaya Sow, 22
Elhadji Abdoul Aziz Sow
Gora Sow, 26
Idrissa Sow, 30
Lamine Sow
Mamadou Sow
Saliou Sow
Samba Sow, 34
Abdou Sy
Alassane Amadou Sy, 25
Aliou Khoudara Sy, 4
Hadji Sy, 18
Hamady Sy
Hawa Sy, 2
Seynabou Sy, 3
Alphousseyni Sy Sylla
Bourama Camara Sylla
Mamadou Sylla, 26
Mamadou Sylla, 31
Mbène Sylla
Saly Sylla, 42
Boury Mberry Sylla
Khady Katy Sylva, 18
Margot Anna Sylva Sylla
Amidou Tall
Mamadou Tall
Papa Tall, 9
Amy Tamba, 41
Chérif Tamba, 39
Fatou Bintou Tamba, 12
Maïmouna Tamba
Youssoupha Tamba, 24
Alain-Amboui Gouyou Tendeng, 52
Ambroise Tendeng
Antoine Alibaa Tendeng, 54
Colette Tendeng
Edouard Tendeng, 44
Françoise Soughessy Tendeng, 21

Henriette Tendeng, 0
Insa Tendeng, 14
Jean-Pierre Adioly Tendeng, 13
Jeannot Tendeng, 37
Julienne Tendeng, 42
LéAndré Tendeng, 22
Léontine Tendeng, 48
Paulette Tendeng, 15
Pauline Tendeng, 15
Rosalie Rosaline Tendeng
Abdoulaye Thiam, 29
Abdourahmane Thiam, 23
Amadou Makhtar Thiam, 16
Babacar Pape Thiam, 24
Bachir Thiam, baby
Bachirou Thiam, 20
Baye Thiam
Boubacar Thiam, 25
Dière Désiré Thiam, 30
Djiby Thiam, 21
Djiby Ndiogou Thiam, 22
Ibra Thiam
Mayébél Thiam, 49
Mbaye Thiam
Moctar Guèye Thiam, 22
Moustapha Racine Thiam, 30
Sidy Seydou Moctar Thiam, 23
Aïssatou Thiané, 20
Ousmane Thiaré, 20
Mady Arfang Thiaré, 35
Ibrahima Niass Thiaw
Seynabou Thiaw, 30
Talla Thiaw Thiane, 35
Alassane Thioune
Madjiguène Tine, 32
Mamadou Tine
Paul Gagne Siry Time, 35
Sénébou Tine, 17
Stanislas Tine, Major
Alioune Tounkara
Alpha Tounkara
Dieynaba Tounkara, 25
El Hadji Samba Cissé Tounkara, 28
Ousmane Tounkara
Hady Tounkara
Oussou Tounkara, 0
Draga Diénéba Diedhiou, 0
El Hadji Famara Diédhiou, 42
Assane Touré, 55
Baba Touré, 25
Bas Bintou Touré, 58
Boubacar Touré, 21
Dieynaba Touré, 33
Enfant Touré, 0
Fanta Mady Touré, 16
Mady Touré, 37
Mame Madia Mady Touré, 47
Moustapha Touré, 21
Youssouf Touré, 3
Youssouph Touré, 32
Diawly Traoré
Assane Traoré
Coumba Traoré
Mady Traoré
Ousseynou Traoré
José Pédro Vaz, 33
Fernando U. Vaz
Niels Vermotte
Claire Verschatse, 20
Assane Wade
Cheikh Wade, 27
Cheikh Oumar Wade, 37
Mamadou Ousmane Wade, 22
Mayacine Wade, 31
Mouhamadou Samba Wade, 26
Moussa Wade, 32
Ousmane Wade
Thierno Wade
Bourama Wacka, 21
Beldo site Racky Wane, 37
Amadou Pathé Wine, 30
Félix Wor Voméno, 39
Meïssa Yade, 40
Illiassou Zoubeiri (Zabeirou)

Index

Italic page references indicate figures.

About the Author

Karen Samantha Barton is a professor of geography, GIS, and sustainability at the University of Northern Colorado. Her work has been supported by Fulbright and the National Endowment for the Humanities-CAORC Senior Fellow programs. Professor Barton is also the recipient of the National Council for Geographic Education Teaching Excellence Award.

www.ingramcontent.com/pod-product-compliance
Ingram Content Group UK Ltd.
Pitfield, Milton Keynes, MK11 3LW, UK
UKHW020841290926
15470UKWH00004B/1472

9 781498 585415